Welding symbols on drawings

Related titles from Woodhead's materials engineering list:

Welded design – theory and practice (ISBN 1 85573 537 7)

A thoroughly practical text, but with sufficient theory to aid understanding of the welding parameters of strength, fatigue and failure, *Welded design* provides specialist information on a topic often omitted from engineering courses. It explains why certain methods are used and gives examples of commonly performed calculations and derivation of data.

Arc welding control (ISBN 1 85573 687 X)

This book examines recent developments in modern arc welding. The first part gives an account of the dynamic behaviour of the arc and its power sources. Part II goes on to describe ways of controlling the welding arc through modern electronics. The third part looks at the prospects of the arc sensor for automatic seam tracking in arc welding. An original method for measuring the welding temperature field using the image colorimetric method is described in Part IV and a detailed account of the recognition method of three-dimensional weld grooves is given in Part V.

Arc welding control is essential reading for researchers, academics, technicians, engineers and other professionals involved in welding automation.

Health and safety in welding and allied processes
Fifth edition (ISBN 1 85573 538 5)

This latest edition has been revised to take into account recent advances in technology and legislative changes. Beginning with a description of the core safety requirements, it goes on to describe special hazards found in the welding environment in terms of their effects and strategies that can be adopted to avoid them. It is an essential resource for welders and their managers.

Details of these books and a complete list of Woodhead's materials engineering titles can be obtained by:

- visiting our web site at www.woodheadpublishing.com
- contacting Customer Services (e-mail: sales@woodhead-publishing.com; fax: +44 (0) 1223 893694; tel.: +44 (0) 1223 891358 ext. 30; address: Woodhead Publishing Ltd, Abington Hall, Abington, Cambridge CB1 6AH, England)

Welding symbols on drawings

E. N. Gregory and A. A. Armstrong

CRC Press
Boca Raton Boston New York Washington, DC

WOODHEAD PUBLISHING LIMITED
Cambridge England

Published by Woodhead Publishing Limited, Abington Hall, Abington
Cambridge CB1 6AH, England
www.woodheadpublishing.com

Published in North America by CRC Press LLC, 2000 Corporate Blvd, NW
Boca Raton, FL 33431, USA

First published 2005, Woodhead Publishing Ltd and CRC Press LLC

British Library Cataloguing in Publication Data
A catalogue record for this book is available from the British Library.

Library of Congress Cataloging in Publication Data
A catalog record for this book is available from the Library of Congress.

Woodhead Publishing ISBN 1-85573-589-X
CRC Press ISBN 0-8493-3591-4
CRC Press order number: WP3591

Typeset by SNP Best-set Typesetter Ltd, Hong Kong
Printed by TJ International, Padstow, Cornwall, England

Contents

	Introduction	vii
	Scope	ix
	Standards referred to in this book	x
	Terms and definitions	xi
1	**The need to specify welds**	**1**
2	**The advantages of symbols**	**3**
3	**Welding symbols 1**	**6**
	Butt/groove welds	6
4	**Welding symbols 2**	**9**
	Fillet and edge welds, backing run or weld, flare groove and bevel welds, and plug or slot weld	9
5	**Welding symbols 3**	**11**
	Spot and seam welds, surfacing, and steep flanked butt welds	11
6	**Location of symbols 1**	**13**
	Butt/groove welds	13
7	**Location of symbols 2**	**14**
	Fillet welds	14
8	**Supplementary symbols**	**16**
	Contours of welds	17
	Convex contour	18
	Concave contour	19
	Toes blended smoothly	19
	Other supplementary symbols	20
	Spacer	20
	Back weld and backing weld	20
	Melt through	21
	Consumable insert	21
	Peripheral welds (weld all round)	22
	Field or site weld	22
	Backing strip or backing	23
9	**Dimensions 1**	**24**
	Butt/groove welds	24
	Partial penetration welds	25
	Groove dimensions	25
	Length of butt/groove welds	27
10	**Dimensions 2**	**28**
	Fillet welds – transverse	28
	Deep penetration welds	29
	Double fillet welds	29
	Unequal leg length fillet welds	30

11	**Dimensions 3**	**31**
	Fillet welds – longitudinal	31
12	**Spot and seam welds**	**34**
	Resistance spot welds	34
	Arc spot welds	35
	Projection welds	35
	Seam welds	36
13	**Stud welds**	**37**
14	**Surfacing**	**39**
	Multiple layers	39
15	**Process identification**	**40**
16	**Non-destructive testing symbols – AWS**	**42**
17	**Exercises**	**44**
	Exercise 1: Flange ended pipe	44
	Exercise 2: Vessel	45
	Exercise 3: Tank	46
	Exercise 4: Beam	47
	Exercise 1 solution	48
	Exercise 2 solution 1	49
	Exercise 2 solution 2	51
	Exercise 3 solution	53
	Exercise 4 solution	55

Introduction

Symbols for indicating welded joints on engineering drawings were originally devised by individual drawing offices to provide more useful information than a simple arrow with the instruction 'weld here'. This practice was obviously unsatisfactory, especially when drawings were passed from one company to another and, to solve this problem, the numerous symbols in existence were rationalised to some extent by countries compiling their own standard specifications for welding symbols.

The American system of symbolisation is the AWS system, formulated by the American Welding Society (AWS). All AWS standards comply with the requirements of the American National Standards Institute (ANSI) and are designated ANSI/AWS. This system became widely used throughout the world, mainly because of the oil industry, and today is used by approximately half the world's welding industry. The rest of the world uses the ISO system, designed by the International Organization for Standardization (ISO). However, a number of countries, particularly those with wide trading links, may use one system in their own country but need to use the other to satisfy the requirements of an overseas customer. Hence the need for a comparison of the two systems.

The British system was standardised in 1933 and the latest of five revisions, published in 1995 as BS EN 22553, is identical to ISO 2553.

For some years an ISO committee has been working on combining the ISO and AWS standards on welding symbols. It is expected that a combined standard will be published in the future which will standardise symbols on a worldwide basis.

It is important to appreciate the purpose of welding symbols, which is mainly to transmit information from the designer to one or more persons along the quality system network. This includes the welding engineer, welding supervisors, welders, inspection personnel and inspectors. In many cases it would be unfair to expect the designer to provide all the information possible from welding symbols without the help of a welding engineer and possibly from other welding and inspection personnel.

The minimum information provided by the designer should consist of the location and types of welds and the sizes and lengths of the fillet welds. The latter will require knowledge of the mechanical properties of the parent metal and the available filler metals. This will be simple for mild steel but more complex for low alloy steels, stainless steels and non-ferrous alloys. A lot of supplementary information can be added to a welding symbol but it may be more convenient and, indeed, useful to include this in a written Welding Procedure Specification (WPS). This procedure is recommended in the ANSI/AWS standard.

It is permissible, therefore, to use a standard on welding symbols for guidance, provided that the drawing indicates at least the locations and sizes of welds, any additional information being provided on a WPS or by detailed notes and drawings.

Scope

This book is an updated version of *Weld symbols on drawings* published in 1982. It describes the application of weld symbols in British/European Standard BS EN 22553, International Standard ISO 2553 and American Standard ANSI/AWS A2.4-98.

For full, authoritative details the standards themselves should be consulted.

References to ISO 2553: 1993 and ANSI/AWS A2.4-98 have been shortened, for convenience, to ISO and AWS where the full reference is not of primary concern and the context makes the abbreviated reference clear. The BS EN 22553 Standard is identical to ISO 2553 so any reference to the ISO standard applies equally to the British standard.

Only the representation of a given weld on a drawing is covered in this book. This does not include the design of the welded joint. The drawings are not necessarily to scale and the weld shapes shown are for the purpose of illustration only and do not represent recommended practice.

Four exercises in the use of welding symbols are included. These will be of particular use to students studying welding technology.

Many thousands of engineering drawings are currently in use which have symbols and methods of representation from superseded standards, e.g. BS 499: Part 2: 1980 or ANSI/AWS 2.4-79. The current European, ISO and American standards are substantially similar but the ANSI/AWS A2.4-98 Standard includes some additional welding symbols and symbols for non-destructive testing. This book includes material to cover the application of these additional symbols. Although symbols in the different standards are similar, the arrows showing locations of welds are different, and these important differences are explained.

ISO 2553 contains very limited information on the representation of brazed or soldered joints. These joints are covered in ANSI/AWS A2.4-98, which contains comprehensive information on this topic.

Standards referred to in this book

ISO 2553 is published by the International Organization for Standardization, 1, rue de Varembé, Case postale 56 CH-1211, Geneva, Switzerland. It was adopted by the UK as a dual British and European standard (BS EN 22553). A summary wall chart (BS 499-C) giving an 'at a glance' view of the symbols, for use in welding workshops, was subsequently issued. It is published by the British Standards Institution, 389 Chiswick High Road, London W4 4AL, UK.

Similarly, the ANSI/AWS standard is issued in both full standard (ANSI/AWS A2.4-98) and summary chart (AWS 2.1-WC, AWS 2.1DC) form. These are published by the American Welding Society, 550 NW Le Jeune Road, Miami, Florida 33126, USA.

Further details on these standards as well as others referred to in the text are given below:

ANSI/AWS A2.4-98	Standard symbols for welding, brazing and nondestructive examination.
ANSI/AWS.A3.0-85	Standard welding terms and definitions.
ANSI/AWS.D1.1-2000	Structural welding code.
AWS 2.1 DC	Welding symbol chart (desk size).
AWS 2.1-WC	Welding symbol chart (wall size).
BS 499-C: 1999	European arc welding symbols – symbolic representation on drawings (wall chart based on BS EN 22553: 1995).
BS 499 Part 1: 1991	Welding terms and symbols. Part 1. Glossary for welding, brazing and thermal cutting.
ISO 2553: 1992 and BS EN 22553: 1995	Welded, brazed and soldered joints – symbolic representation on drawings.
ISO 9692-1: 2003 and BS EN 29692-1: 2003	Welding and allied processes – recommendations for joint preparation – manual metal arc welding, gas shielded metal arc welding, TIG welding and beam welding of steels.
ISO 4063: 1990 and BS EN 24063: 1992	Welding, brazing, soldering and braze welding. Nomenclature of processes and reference numbers for symbolic representation on drawings.

Terms and definitions

Some terms and definitions are used throughout this book to clarify the meaning of the surrounding text. In this section they are listed for easy reference along with other relevant terms and definitions.

Comprehensive definitions are included in BS 499: Part 1: 1991 Welding terms and symbols: Glossary for welding, brazing and thermal cutting, and ANSI/AWS A3.0-85: Standard welding terms and definitions.

It is important to note the differences in the usage of certain terms and units in the British (BS), European (EN), International (ISO) and American (AWS) standards. In regard to the terms **butt weld** and **groove weld**, British, European and International systems use 'butt weld', whereas the American system uses 'groove weld'.

The terms **weld symbol** and **welding symbol** are not defined in BS 499: Part 1: 1991 Welding terms and symbols or in ANSI/AWS A3.0-85: Standard welding terms and definitions. They are explained, however, in AWS/ANSI A2.4–98 Standard symbols for welding, brazing, and non-destructive examination.

In the UK 'weld symbol' and 'welding symbol' are interchangeable by common usage but in the American standard on symbols they have different meanings. 'Weld symbol' is the basic V, U or triangle, representing, respectively, single-V, single-U or fillet welds. 'Welding symbol' means a reference line to which the weld symbols can be added and an arrow line pointing to the position of the welded joint. Additional elements may be added such as weld sizes and lengths, welding process and non-destructive testing requirements, which all contribute to the welding symbol.

In regard to **US customary**, **imperial** and **metric units**, ANSI/AWS A2.4-98 requires that the system (US, imperial or metric) used for the dimensions on a drawing shall also be used as part of the welding symbol. In the ISO system, which uses metric units, weld dimensions can be written adjacent to the symbols.

Actual throat thickness	The shortest distance between the weld root and the face of a fillet weld (*see* Fig. 10.3 on page 28).
Arc spot weld	A spot weld made by an arc welding process.
Back bead	A weld bead resulting from a back weld pass.
Back gouging	Removal of weld and parent metal from the other side of a partially welded joint to facilitate fusion and complete joint penetration following welding from that side.
Back weld	A weld made at the back of a single groove weld.
Backing	Material or device placed against the back of the joint to support and retain molten weld metal. The backing may be either permanent or temporary.
Backing pass	A weld pass made for a backing weld.
Backing weld	Backing in the form of a weld.
Bevel angle	The angle at which the edge of a component is prepared for making a weld.

Butt/groove weld	A weld made to join two members aligned in the same plane (*see* butt, T- and corner joints. *See also* explanation of the difference between ISO and AWS nomenclature on page xi.)
Butt joint	A joint between two members aligned approximately in the same plane.
Butt weld	*See* butt/groove weld.
Corner joint	A joint between two members located approximately at right angles to each other (in the shape of an 'L').
Cruciform joint	A joint in which two flat plates are welded to another flat plate at right angles and on the same axis.
Design throat thickness	*See* effective throat thickness.
Double bevel butt/groove weld	A butt/groove weld in the joint preparation for which the edge of one component is double bevelled and the fusion face of the other component is at right angles to the surfaces of the first component.
Double-J butt/groove weld	A butt/groove weld in the joint preparation for which the edge of one part is prepared so that in cross-section the fusion face is in the form of two opposing 'J's and the fusion face of the other part is at right angles to the surfaces of the first component.
Double-U butt/groove weld	A butt/groove weld in the joint preparation for which the edges of both components are shaped so that in cross-section the faces form two opposing 'U's having a common base.
Double-V butt/groove weld	A butt/groove weld in the joint preparation for which the edges of both components are bevelled so that in cross-section the fusion faces form two opposing 'V's.
Double welded joint	A joint that is welded from both sides.
Edge preparation	The surface prepared on the edge of a component to be welded.
Edge weld	A weld on an edge joint used for joining two or more parts and in which the weld metal covers part or the whole of the edge widths.
Effective throat thickness	The minimum distance minus any convexity between the weld root and the face of a fillet weld (*see* Fig. 10.3 on page 28). Also called design throat thickness.
Faying surface	The mating surface of a member that is in contact with or very close to another member to which it is to be joined.
Fillet weld	A fusion weld, other than a butt edge or fusion spot weld, that is approximately triangular in transverse cross-section.
Fusion face	The portion of a surface or of an edge that is to be fused in making a fusion weld.
Groove weld	*See* butt/groove weld.
Heat affected zone (HAZ)	The part of the parent metal that is metallurgically affected by the heat of welding but not melted.
Leg length	The distance of the actual or projected intersection of the fusion faces and the toe of a fillet weld, measured across the fusion face (*see* Fig. 10.1 on page 28).
Nominal throat thickness	The perpendicular distance between two lines, the one drawn through the outer toes and the other through the deepest point of fusion penetration.
Partial penetration	Penetration that is less than complete.
Penetration	The depth to which the parent metal has been fused.
Plug weld	A weld made by filling a hole in one component of a workpiece with the filler metal so as to join it to the surface of an overlapping component exposed through the hole.
Projection weld	A weld made by a resistance welding process in which the localizing of force and current to make the weld is obtained by the com-

	ponent shape or by the use of a projection or projections raised on one of the faying surfaces.
Root face	The portion of a fusion face at the root that is not bevelled or grooved.
Root gap	A separation at the joint root between the workpieces.
Seam weld (resistance)	A weld made by a resistance welding process in which force is applied continuously and current continuously or intermittently to produce a linear weld.
Single bevel butt/groove weld	A butt/groove weld in the joint preparation for which the edge of one component is bevelled and the fusion face of the other component is at right angles to the surfaces of the first component.
Single-J butt/groove weld	A butt/groove weld in the joint preparation for which the edge of one component is shaped so that in cross-section the fusion face is in the form of a 'J' and the fusion face of the other part is at right angles to the surfaces of the first component.
Single-U butt/groove weld	A butt/groove weld in the joint preparation for which the edges of both components are shaped so that in cross-section the faces form a 'U'.
Single-V butt/groove weld	A butt/groove weld in the joint preparation for which the edges of both components are bevelled so that in cross-section the fusion faces form a 'V'.
Single welded joint	A joint that is welded from one side only.
Slot weld	A weld joining two overlapping components and made by depositing a fillet weld round the periphery of a hole in one component.
Spot weld (resistance)	A weld made by a resistance welding process that produces a weld at the faying surfaces of a joint by the heat obtained by resistance to electric current from electrodes which concentrate the current and pressure at the weld area.
Square butt/groove weld	A butt/groove weld in the joint preparation for which the fusion faces lie approximately at right angles to the surfaces of the components to be joined and are substantially parallel to one another.
Stud weld	A weld joining a metal stud to a workpiece, the weld being made over the whole end area of the stud. Welding may be by arc, resistance or other suitable process, with or without external gas shielding.
Supplementary symbol	A symbol added to the basic welding symbol, providing further information.
Surfacing	The application of material to the surface of a component by welding, brazing or spraying to increase wear resistance or corrosion resistance.
T-joint	A joint between two members located approximately at right angles to each other in the form of a 'T'.
Throat thickness	*See* actual throat thickness, design throat thickness (effective throat thickness) and nominal throat thickness.
Weld symbol	*See* welding symbol.
Welding Procedure Specification (WPS)	A document that has been qualified by an approved method and provides the required variables of the welding procedure to ensure repeatability during production welding.
Welding symbol	A diagrammatic or pictorial representation of the fundamental characteristics of a weld. *See* explanation of the difference between ISO and AWS nomenclature on page xi.

1 The need to specify welds

It is sometimes argued that it is unnecessary to specify welds on drawings and that the welder should be relied upon to deposit a suitable weld. This practice can be extremely risky because the type and size of the weld must be appropriate for the parent material and service conditions of the fabrication, and the necessary information and data are normally available only in the design office.

Figure 1.1 illustrates **(a)** the instruction 'weld here' and **(b–d)** three ways to follow this instruction.

The instruction 'weld here', illustrated in **Fig. 1.1(a)**, is rarely seen on a drawing because it is open to a number of different interpretations as shown in Fig. 1.1(b), (c) and (d).

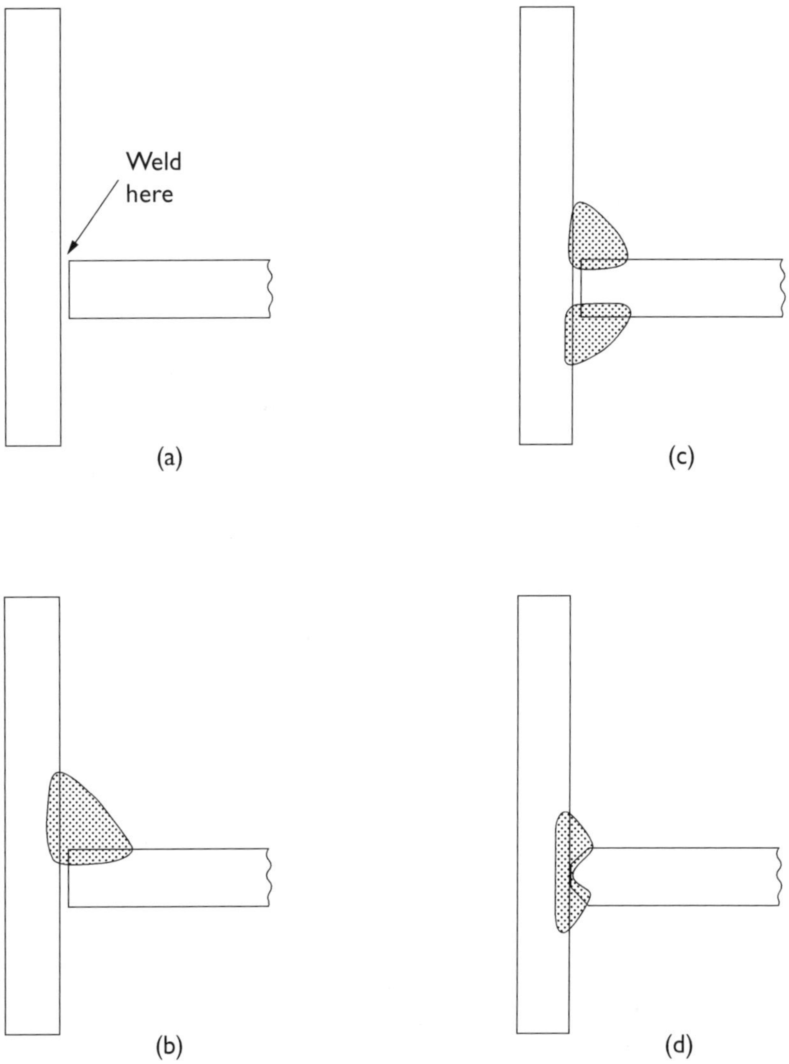

1.1 (a) The instruction 'weld here' and (b–d) three ways to follow this instruction.

Figure 1.1(b) shows a single fillet weld. This weld is simple and therefore cheap to apply but could be seriously deficient in performance.

Figure 1.1(c) shows a double fillet weld, which takes longer to apply. Unless access is available to both sides of the joint, it will be impossible to weld it.

Figure 1.1(d) illustrates a T-butt/groove weld. This weld normally requires edge preparation on a horizontal member, and therefore is more complex and expensive. However, it may be essential for certain service conditions.

It can be seen from the previous examples that major problems will arise unless welded joints are carefully specified by the design office. The situation is particularly critical where, for example, work is placed with a subcontractor and the instructions need to be especially precise.

2 The advantages of symbols

When it is required to indicate a weld on a drawing, it may seem that the weld can simply be drawn as it will appear. In the majority of cases, symbolic representation can be used to cut down the time needed to complete the drawing and improve clarity.

To save time in drawing the edge preparation for a butt/groove weld or the shape and size of a fillet weld, a set of weld symbols can be used. These symbols are placed on a horizontal reference line. This line is attached to an arrow line which points to the location of the weld (see **Fig. 2.1**). In the ISO system there are two parallel reference lines, one solid and one dashed. In the AWS system a solid reference line is used.

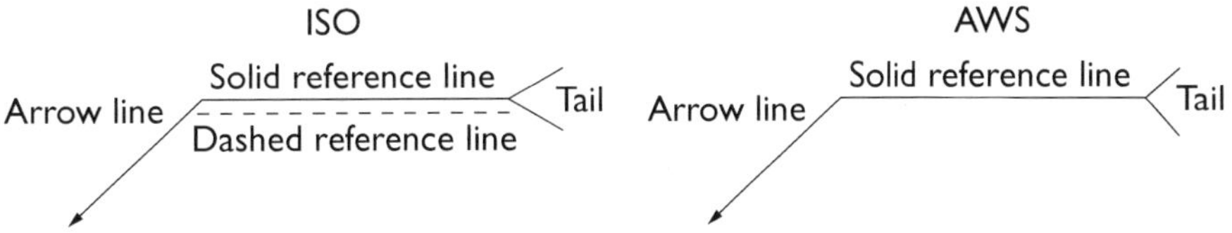

2.1 ISO and AWS reference lines and arrow lines.

Apart from weld symbols placed on the reference line, additional information can be supplied adjacent to the tail which is generally omitted when not required.

The arrow line can point in any direction as shown in **Fig. 2.2**. This is so that it can locate welds in any welding position, for example flat or overhead. The arrow line is never drawn horizontally because this would make it appear to be a continuation of the reference line, which is always horizontal.

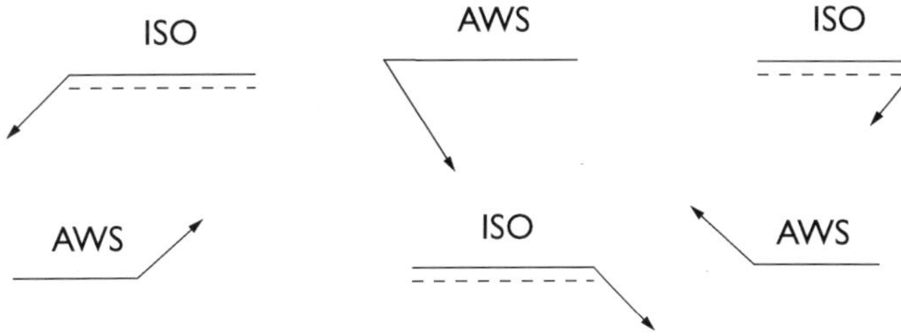

2.2 Possible directions in which arrow lines may point.

It is conventional practice to refer to the opposite sides of a welded joint as the arrow side and the other side (see **Fig. 2.3**).

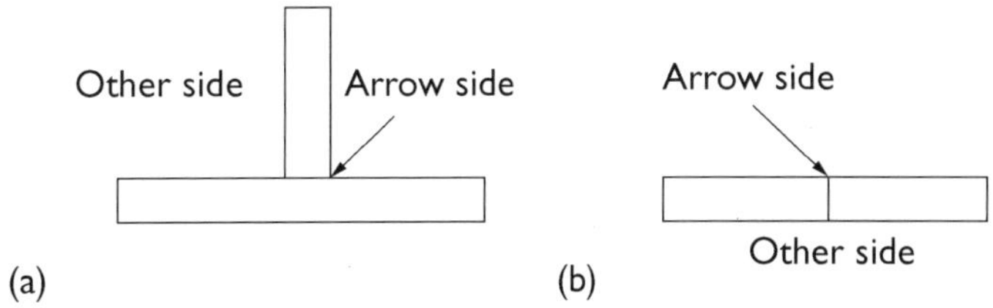

2.3 The arrow side and other side of a T-joint (a) and a butt/groove joint (b).

In the ISO system a weld on the arrow side is indicated by placing the weld symbol on the solid reference line and a weld on the other side has the symbol on the dashed line, as shown in **Fig. 2.4**.

In the AWS system the weld symbol for a weld on the arrow side is placed below the line and for a weld on the other side the symbol is placed above the line.

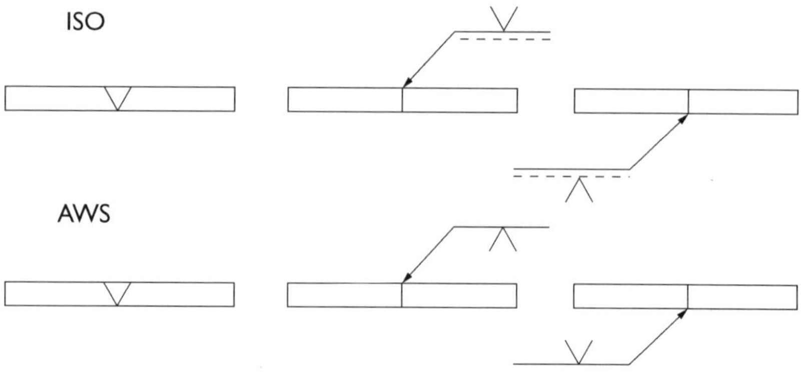

2.4 Symbolisation of a weld on the arrow side and a weld on the other side in the ISO and AWS systems.

In the ISO system the dashed line can be drawn above or below the solid line but the symbols on the solid line always refer to the arrow side of the joint. Symbols on the dashed line indicate a weld on the other side. It is recommended that the solid line is always drawn above the dashed line as standard practice. If a weld is made on both sides, as in a double fillet weld, the weld symbol is placed on both sides of the reference line or lines, in which case, in the ISO system, the dashed line can be omitted.

Figure 2.5(a–f) shows the use of symbols to indicate the type and size of a T-butt weld and a double fillet weld.

Without using weld symbols, **Fig. 2.5(a)** shows a drawing of a T-butt weld with 6 mm leg length fillet welds and with the edge preparation shown.

Figure 2.5(b) shows the T-butt weld represented by symbols which convey all the information according to the ISO standard.

Figure 2.5(c) shows the T-butt weld indicated by weld symbols for the AWS standard.

If the section thicknesses of parts are small compared with their overall size, as in box girders, the welds are often too small to be drawn to scale and to be reproduced accurately, as illustrated in **Fig. 2.5(d)** for a double fillet weld.

Figure 2.5(e) and **(f)** shows the double fillet weld indicated by ISO and AWS symbols.

In order to simplify drawings as much as possible, the ISO standard recommends that, where appropriate, full details of edge preparations should be shown separately. This is also in accordance with AWS recommendations.

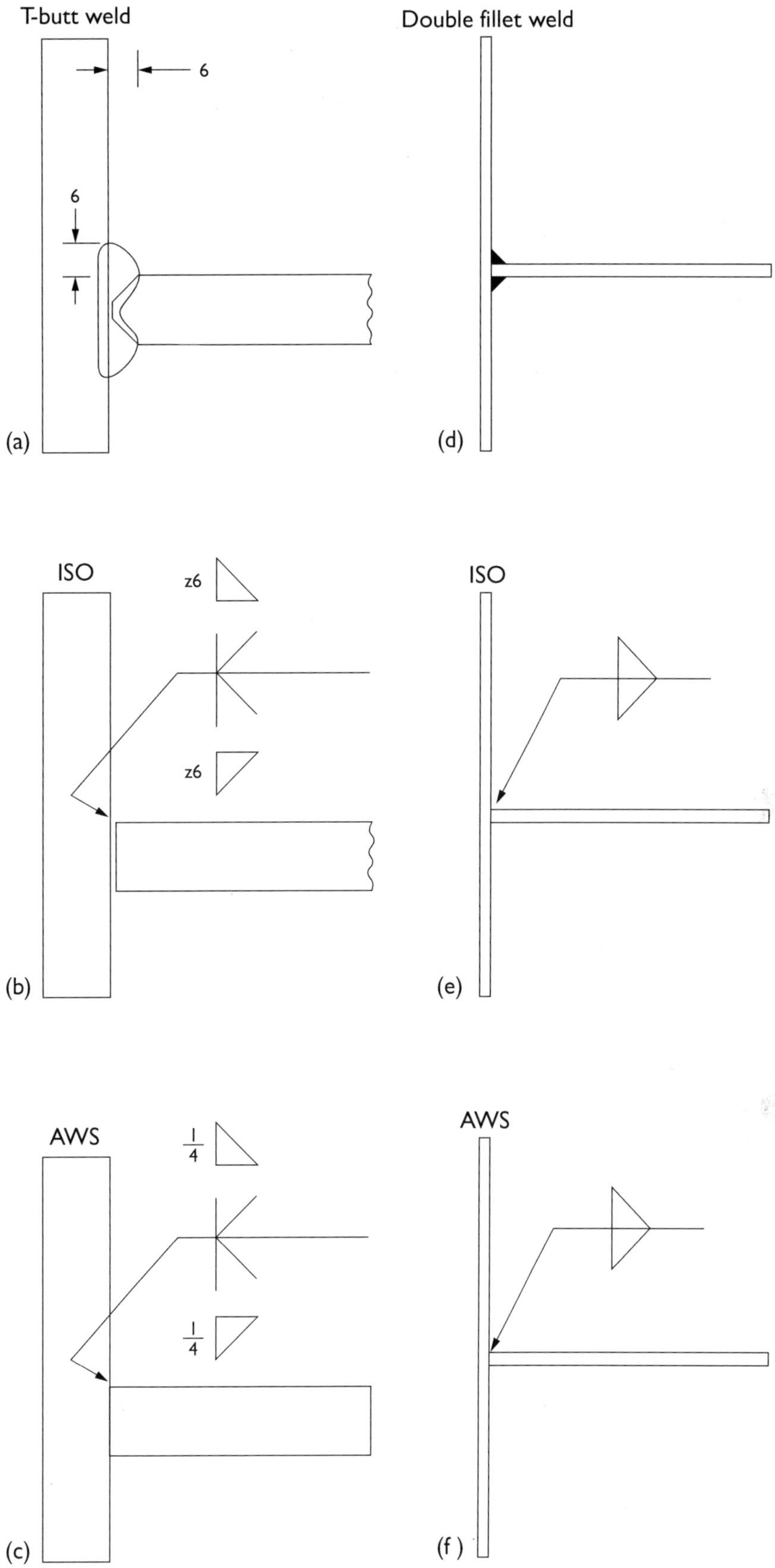

2.5 (a–f) Location of symbols in a T-butt weld and a double fillet weld. ISO dimensions are in millimetres; AWS dimensions are in inches.

3 Welding symbols I

Butt/groove welds

The butt/groove welding symbols are shown in **Fig. 3.1(a–h)**.

Figure 3.1(a) illustrates a **single-V butt/groove weld**, which is the commonest form of edge preparation for this type of weld.

Figure 3.1(b) shows a **square butt/groove weld**. This weld will be limited to a maximum section thickness depending on the welding process used. If a backing strip is used, the section thickness can be increased considerably.

Guidance on edge preparations is included in BS EN 29692 and ISO 9692: 1992, in which the range of thickness recommended for this type of weld is 3–8 mm.

Without a backing strip, a maximum section thickness of 4 mm is recommended with a gap equal to the thickness. With a backing strip, a gap of 6–8 mm is recommended. Dimensions of edge preparations are not included with weld symbols in ISO 2553 but these can be included with AWS symbols. This can make a drawing complex and, in some cases, may lead to confusion. It is preferable to include details of edge preparations in a Welding Procedure Specification (WPS).

Figure 3.1(c) shows a **single bevel butt/groove weld**. This edge preparation is generally used when it is only possible to prepare one edge of adjoining sections.

Figure 3.1(d) illustrates a **single-U butt/groove weld**, which is used to restrict the quantity of weld metal required in sections greater than 12 mm thick.

Figure 3.1(e) shows a **single-J butt/groove weld**. This weld is used to restrict the quantity of weld metal required in sections greater than 16 mm thick when it is only possible to prepare one edge of adjoining sections.

Figure 3.1(f) illustrates a **butt weld between plates with raised edges** (ISO) or **edge weld on a flanged groove joint** (AWS). The AWS term is a more accurate description of this weld, which is an edge weld, described in Section 4. In this weld the edges are melted down to form a low strength sealing weld. In the AWS system, if full penetration is required, the welding symbol includes the melt-through symbol (shown in Fig. 8.10 on page 21) placed on the opposite side of the reference line.

Figures 3.1(g) and **(h)** indicate **single-V** and **single bevel butt welds with broad root faces**. These symbols are included in ISO 2553 but not in AWS A2.4–98. They are illustrated in Fig. 3.1 but best avoided, as described below.

Dimensions of a broad root face are specified in ISO 9692: 1992. A root face of 2–3 mm is specified for section thicknesses of 5–40 mm, whereas for a single-V butt weld (Fig. 3.1(a)) a maximum root face of 2 mm is used for thicknesses of 3–10 mm. This is unnecessarily confusing and it is recommended that the broad root face

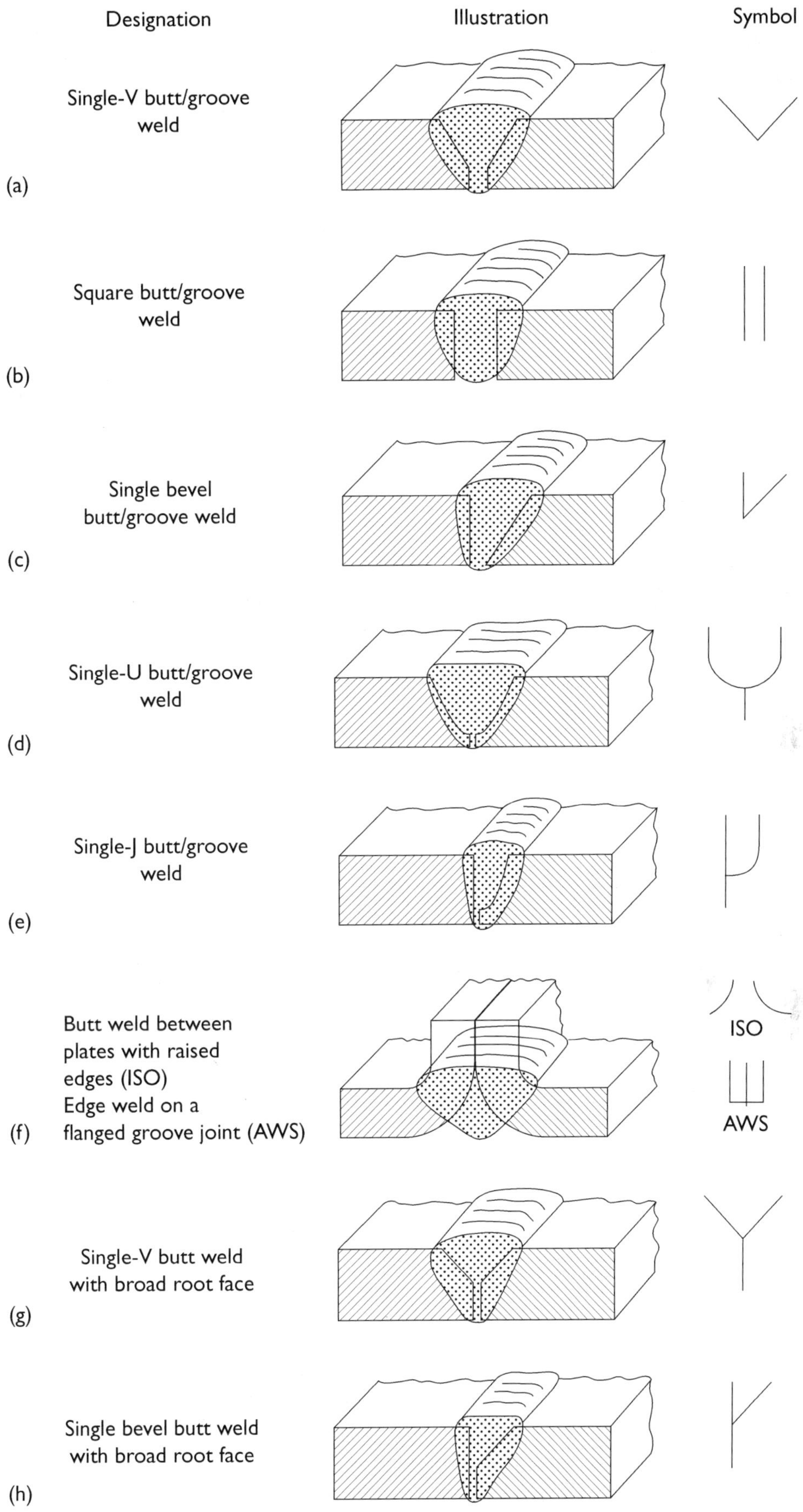

3.1 (a–h) Examples of elementary welding symbols – 1.

terms and symbols should be avoided. A root dimension can be specified whatever its thickness and there is no point in having a special definition 'broad root face' when the root face is greater than 2 mm.

As stated previously, when ISO 2553 is used, the dimensions of the edge preparations are not included as part of the welding symbol and should be given as part of the WPS. With the AWS system, the depth of the groove can be specified by a number on the left hand side of the weld symbol. This dimension, subtracted from the section thickness, will indicate the size of the root face (**Fig. 3.2**).

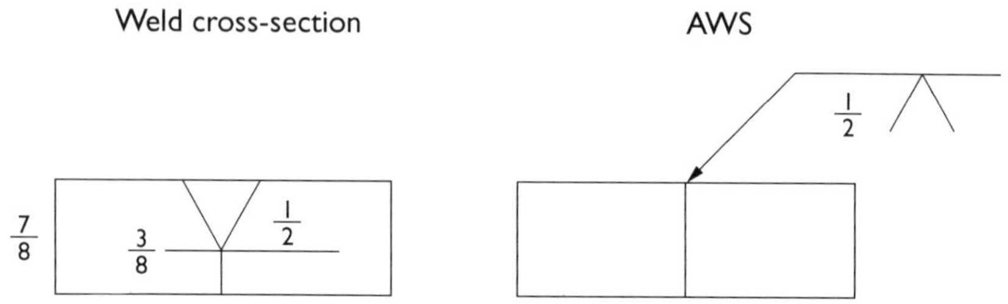

3.2 Size of root face (dimensions are in inches).

4 Welding symbols 2

Fillet and edge welds, backing run or weld, flare groove and bevel welds, and plug or slot weld

The symbols for fillet and edge welds, backing run or weld, flare groove and bevel, and plug or slot welds are shown in **Fig. 4.1(a–f)**.

Figure 4.1(a) illustrates a **fillet weld**. Unless otherwise indicated, the leg lengths are normally equal.

Figure 4.1(b) shows an **edge weld**. The ISO and AWS symbols are fairly similar and are drawn above and below the reference line respectively, both indicating a weld on the arrow side. However, there is no possibility of confusion because the edge weld can only be deposited on one side.

Figure 4.1(c) shows a **backing run or weld**. This is not a weld in its own right, as this symbol is not used on its own. It is deposited on the opposite side of the joint to the main weld, so both sides must be accessible.

The AWS standard includes symbols for flare-V-groove and flare-bevel-groove welds. A **flare-V-groove weld**, shown in **Fig. 4.1(d)**, is a weld in a groove formed by two members with curved surfaces. A **flare-bevel-groove weld**, shown in **Fig. 4.1(e)**, is a weld in a groove formed by a member with a curved surface in contact with a planar member. The commonest application for these welds is in the welding of reinforcing bars.

Figure 4.1(f) shows a **plug or slot weld**, which is a circular or elongated hole completely filled with weld metal. The size of the hole should be restricted to avoid excessive distortion and unnecessary consumption of filler metal.

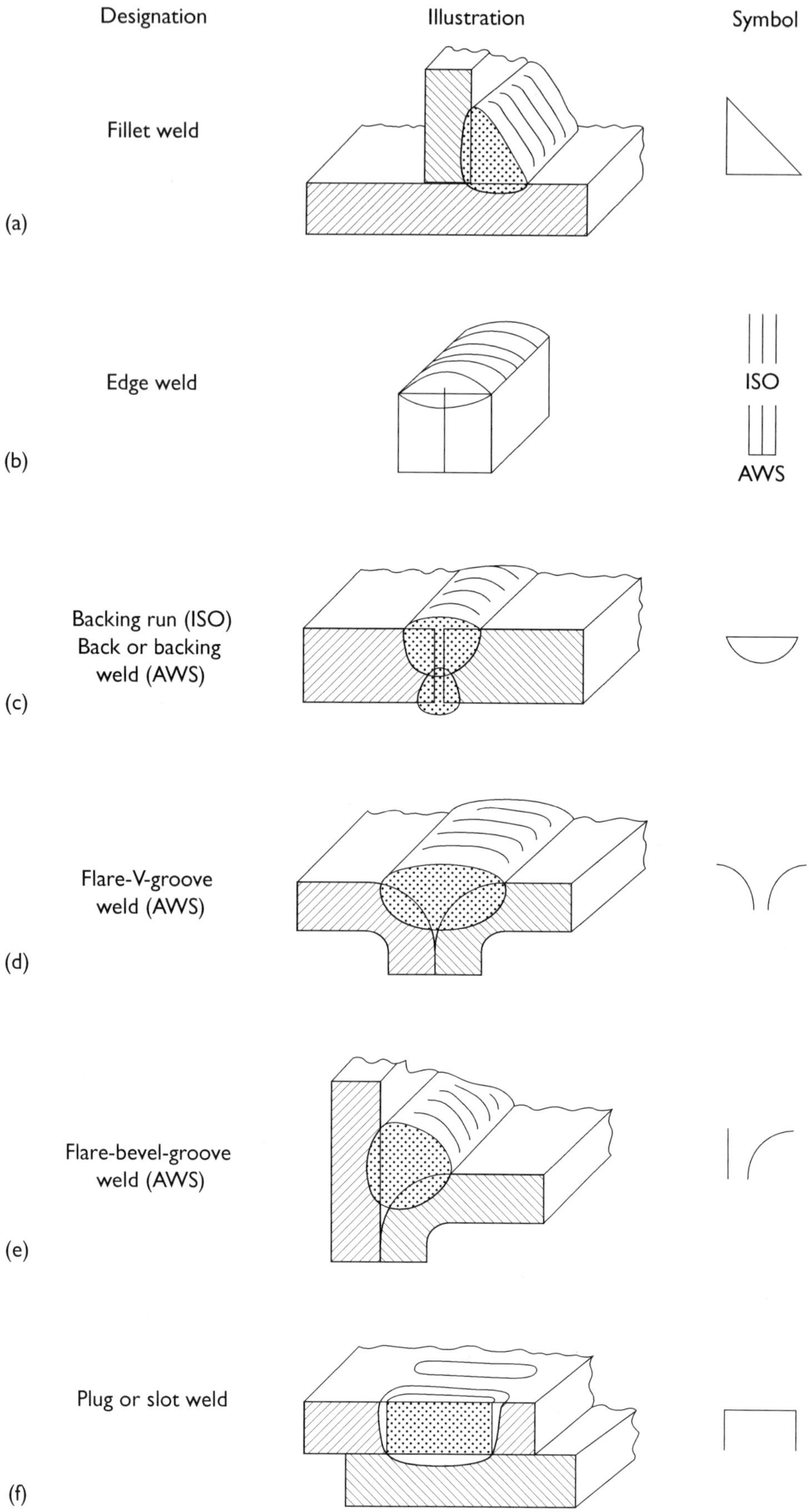

4.1 (a–f) Examples of elementary welding symbols – 2.

5 Welding symbols 3

Spot and seam welds, surfacing, and steep flanked butt welds

The symbols for resistance and arc spot and seam welds are shown with reference lines (ISO) to indicate clearly the position of the symbols in relation to the line. AWS symbols would be similarly placed on the reference line for resistance welds and below the line for arc welds. These symbols are shown in **Fig. 5.1(a)** and **(b)**.

Figure 5.1(a) shows **spot welds**. The upper illustration shows a resistance spot weld or projection weld requiring access from both sides. The lower illustration shows an arc spot weld made from one side of the joint. The reference line is on one side of the symbol.

Figure 5.1(b) illustrates **seam welds**. The upper illustration shows a resistance seam weld requiring access from both sides of the joint. The lower illustration shows an arc seam weld made from one side of the joint.

Arc spot and arc seam welding processes are rarely used and, by common usage, spot and seam welding mean resistance welding.

Figure 5.1(c) indicates **surfacing**. In this symbol, the arrow line points to the surface to be coated with weld metal.

ISO 2553 does not explain how to indicate the extent of the surface coating which is essential information.

AWS A2.4-98 gives detailed instructions on how to show the area to be coated on a plan view; this is described in Section 14.

Figure 5.1(d) shows **steep flanked butt welds**. ISO 2553 includes two symbols representing a steep flanked single-V butt weld and a steep flanked single-bevel butt weld. The edge preparations for these welds are shown in ISO 9692-1: 2003 with bevel angles of 5–20° for a butt weld and 15–30° for a bevel butt weld. The welds have a backing strip.

These symbols are not included in the AWS standard and are not really necessary because the welds are, in fact, single-V and single-bevel butt welds.

They can be indicated as such by including the symbol for a backing strip (shown in Fig. 8.1 on page 16). Both symbols in Fig. 5.1(d) have a horizontal line at the bottom. The lines vary slightly in length. This difference is pointless because, as in the case of the fillet weld symbol, the symbols would be placed with the lines directly on the reference line.

There are no examples of the application of these symbols in ISO 2553.

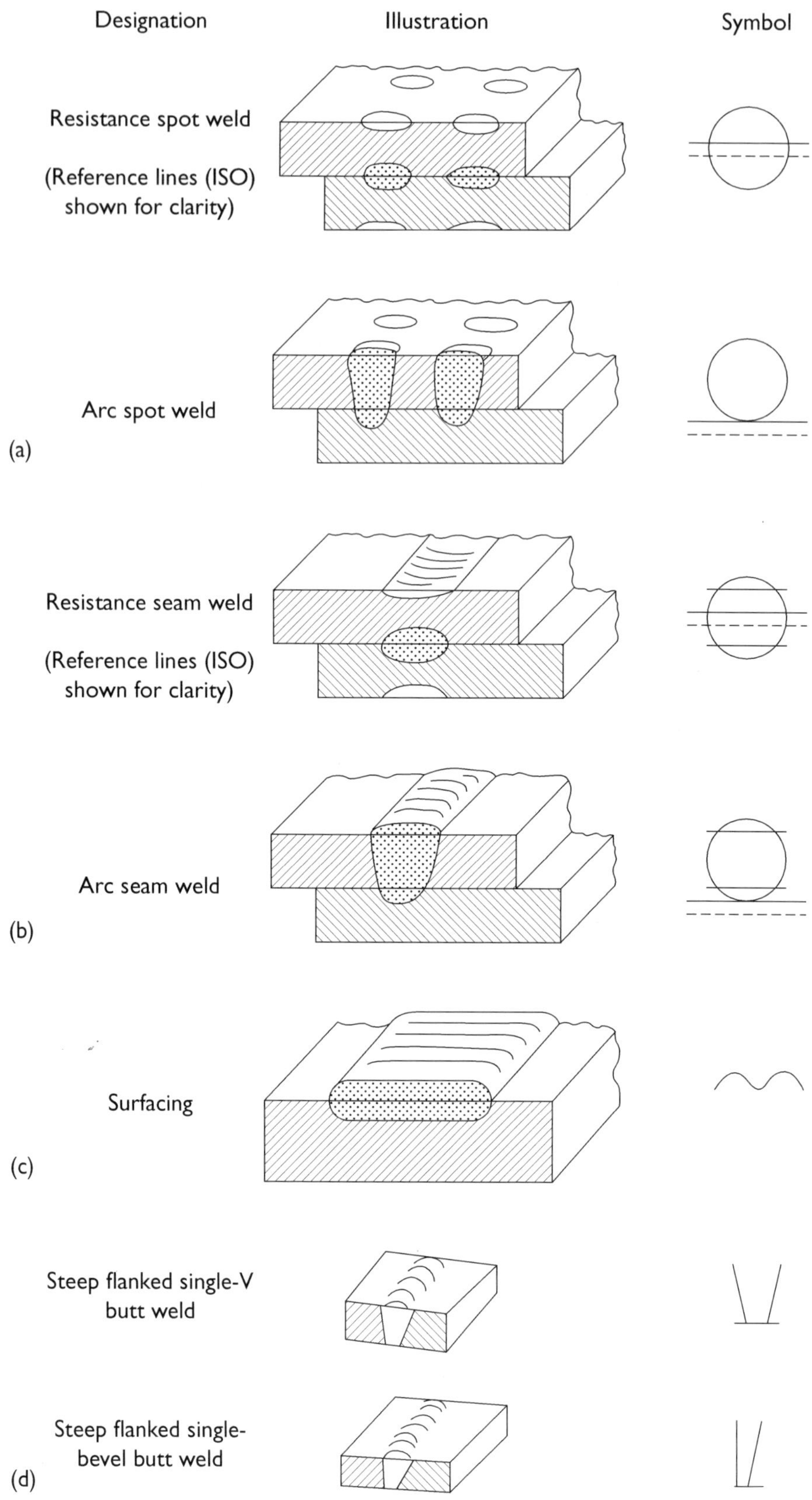

5.1 (a–d) Examples of elementary welding symbols – 3.

6 Location of symbols I

Butt/groove welds

Figure 6.1 (a–c) shows the location of butt/groove welding symbols. For the single-V butt weld shown in **Fig. 6.1(a)**, the welding symbols are located on the reference line which is connected to an arrow pointing to one side of the joint. The arrow can point to the weld in a plan view, as shown in **Fig. 6.1(a)**, or a cross-section, as shown in **Fig. 6.1(b)**.

Figure 6.1(c) shows a single-bevel butt/groove weld in which the arrow line points to the edge of the joint which is to be prepared with a bevel.

The AWS standard specifies that when only one edge of a joint is to be prepared, as in a single-bevel or J-groove weld, the arrow line should be drawn with a break (more accurately described as a sharp bend) as shown in Fig. 6.1(c), with the arrow pointing to the prepared edge. The arrow line need not be bent if it is obvious which edge of the joint is to be bevelled or if there is no preference as to which edge is to be prepared.

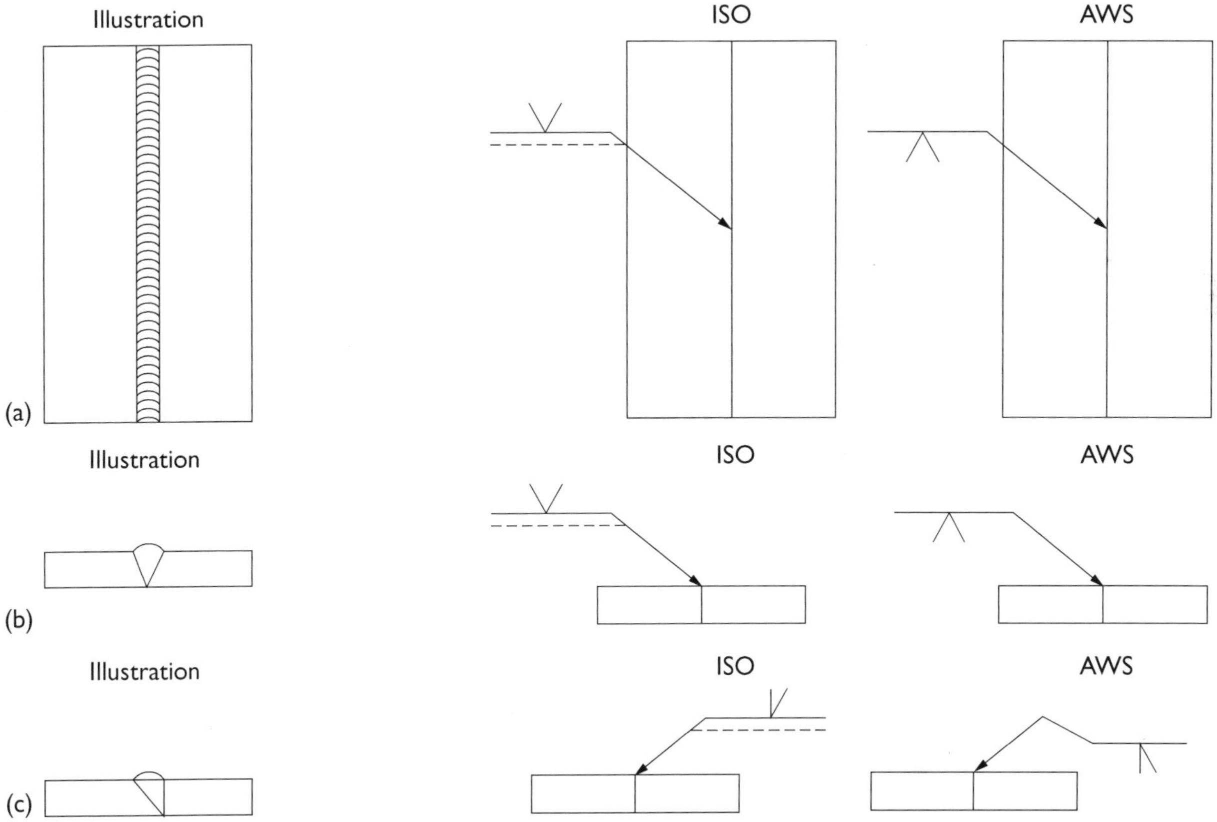

6.1 (a–c) Location of butt/groove welding symbols.

7 Location of symbols 2

Fillet welds

As with butt welds, weld symbols for fillet welds are located on a reference line connected to an arrow which points to one side of the joint. In the ISO system the symbol for a weld on the arrow side is placed on the continuous line and the symbol for a weld on the other side is placed on the dashed line. In the AWS system the symbol for a weld on the arrow side is placed below the single continuous line and the symbol for a weld on the other side is placed above the line. This is illustrated in **Fig. 7.1(a–d)** for a T-joint (a joint between two members, located approximately at right angles to each other to form a 'T') and a cruciform joint (a joint in which two flat plates are welded to another flat plate at right angles and on the same axis).

Figure 7.1(a) shows an end view of a T-joint with a single fillet weld. The shape of the weld would not normally be shown on an engineering drawing.

Figure 7.1(b) shows an end view of a cruciform joint. The two welds are on different joints, i.e. they do not form a double fillet weld. Therefore, two separate arrows are required to indicate two single fillet welds.

In **Fig. 7.1(c)** there is a double fillet weld on the left of the section and a single fillet weld on the right-hand side. The fillet weld symbol is always drawn with the upright leg on the left.

For the joint in **Fig. 7.1(d)**, the need to show two symbols, one on each side of a vertical member, can be avoided by the use of more than one arrow line. This practice is not specifically authorised in ISO 2553 but in AWS A2.4-98 it is stated that two or more arrows may be used with a single reference line to point to locations where identical welds are specified.

This practice should be used with caution to avoid a drawing with a minimum of weld symbols and a multitude of arrow lines criss-crossing the drawing.

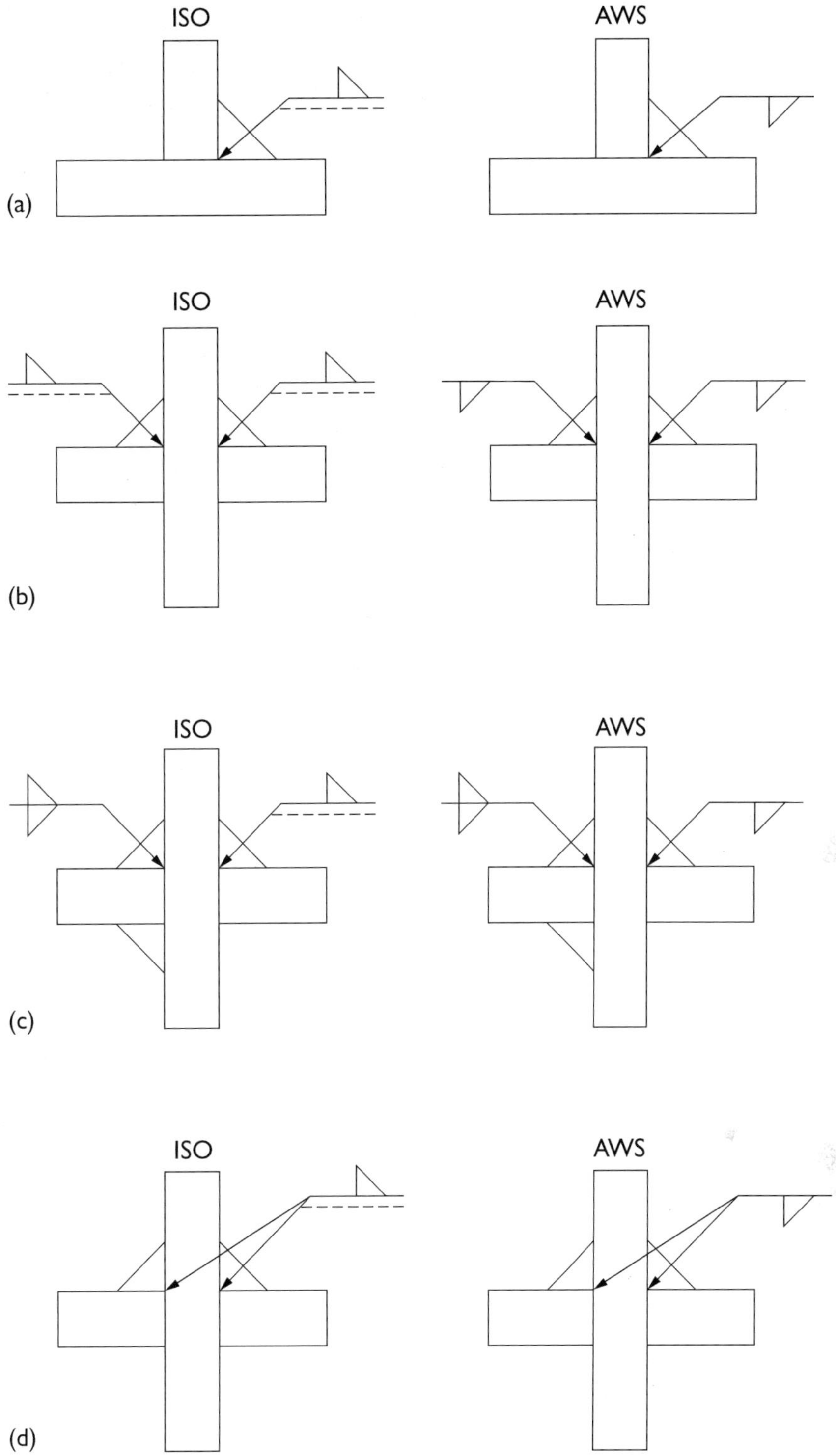

7.1 (a–d) Location of fillet welds.

8 Supplementary symbols

Additional information about a weld can be provided by supplementary symbols used in conjunction with those welds already described. In most cases, the same symbol is used in the ISO and AWS standards. In other cases, only one of the standards uses a symbol for a particular requirement. A comparison of symbols used in the ISO and AWS standards is shown in **Fig. 8.1**.

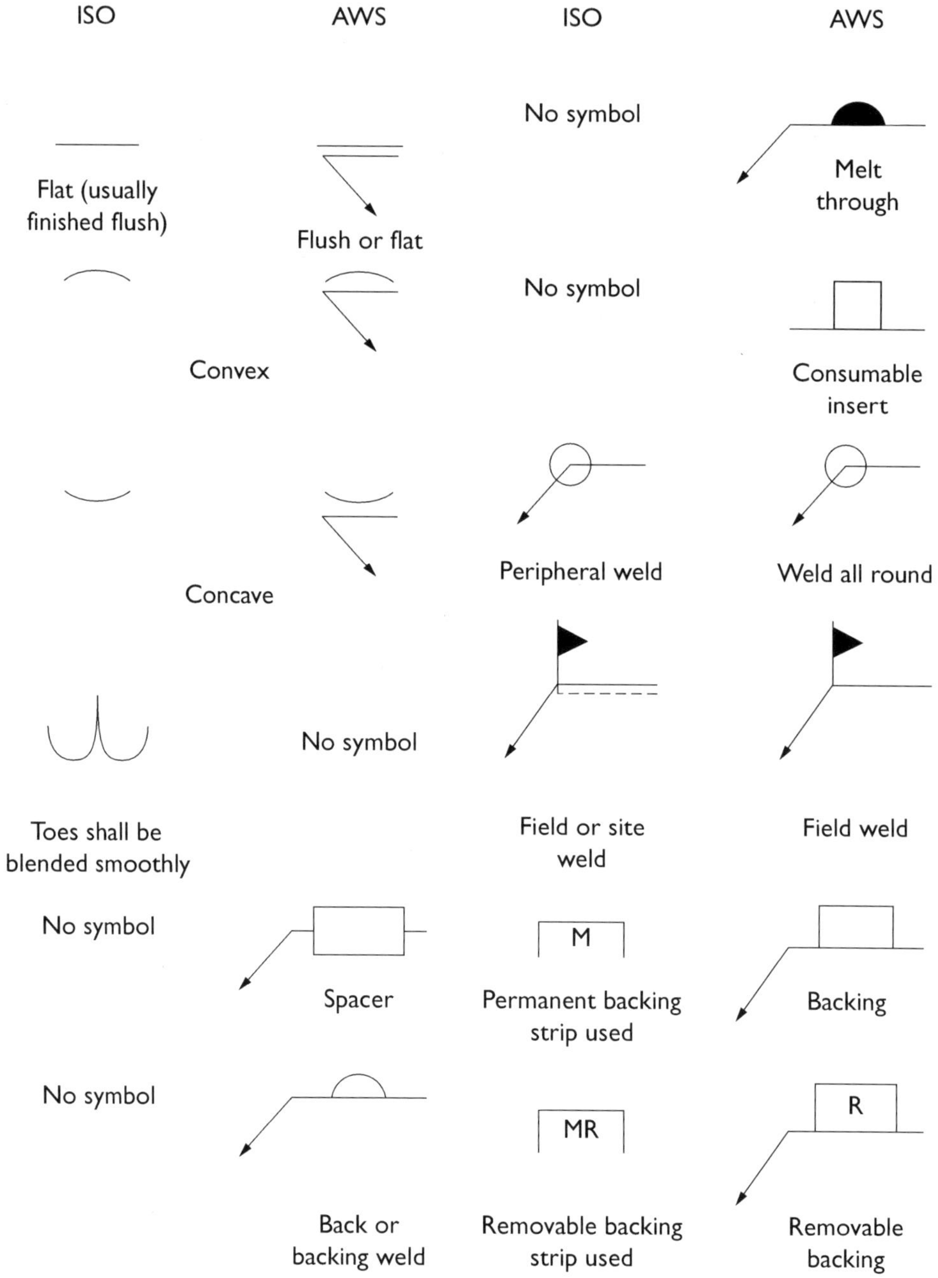

8.1 Supplementary symbols.

The supplementary symbols in AWS A2.4-98 are all shown with a reference line and an arrow line which are not actually part of the welding symbol. All the symbols in Fig. 8.1 are drawn as they appear in the standards. Their application is shown in the following pages.

Contours of welds

Figure 8.1 includes the symbols that are used to indicate the required shape of a weld. The AWS standard states that welds to be made with a flush, flat, convex or concave contour, without the use of mechanical finishing, shall be specified by adding the flush or flat, convex or concave symbol to the welding symbol.

This practice seems slightly pedantic and it is questionable to what extent it is carried out.

Figure 8.2 shows the cross-section of a single-V butt/groove weld with the weld face flat and flush with the plate surface.

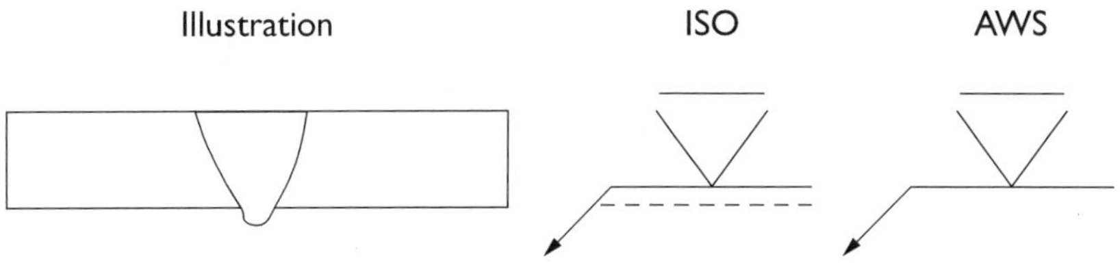

8.2 Flat or flush contour – 1.

In the ISO system it is likely that the Welding Procedure Specification (WPS) would contain instructions on post-weld finishing treatment required, such as grinding or machining.

In contrast to this, in the AWS system the symbol indicates that the surface finish is to be achieved in the as-welded condition.

Figure 8.3 shows a single-V butt/groove weld with the weld face and the penetration bead flush with the plate surface.

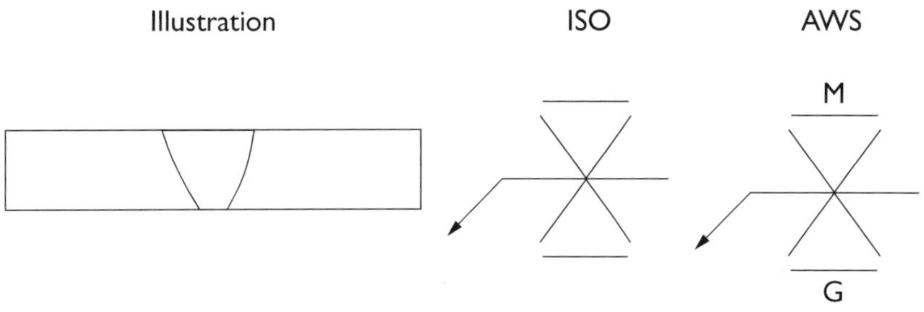

8.3 Flat or flush contour – 2.

In the AWS system the method of post-weld finishing is shown by the capital letters M and G, indicating machining and grinding. Other letters used in the AWS standard are:

C – chipping
H – hammering
R – rolling.

If the weld is to be finished by an unspecified mechanical means the letter U is used in the AWS system, as shown in **Fig. 8.4**.

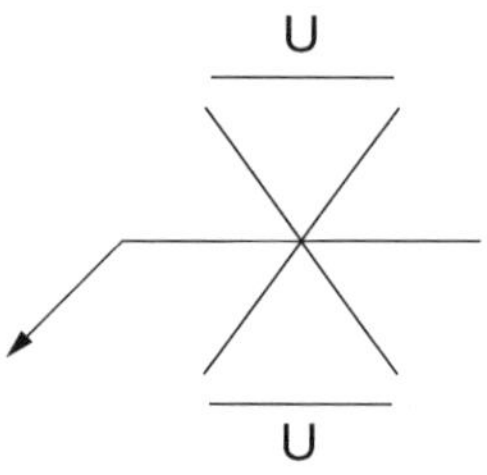

8.4 Unspecified finishing (AWS).

Convex contour

The symbol for a convex weld ⌒ is rather mysterious because it is difficult to imagine a case where a convex profile would be specified for either a butt/groove weld or a fillet weld. All welds are normally deposited with a slightly convex profile to provide the minimum required throat thickness without excess weld metal. This symbol, without further instructions, could cause confusion in a welding shop because it would encourage a welder to deposit excess weld metal, which would create potential problems and additional costs.

The location of the convex weld symbol is shown in **Fig. 8.5**.

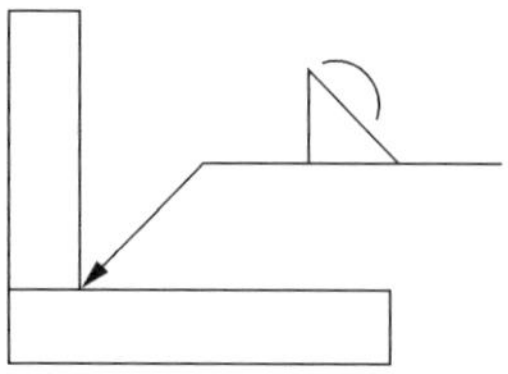

8.5 Location of the convex weld symbol.

It is worth noting that AWS D1.1-2000 Structural Welding Code specifies limits to convexity of welds depending on the width of the weld face as follows:

Width of weld face	*Maximum convexity*
W ≤ 8 mm	2 mm
W > 8 to < 25 mm	3 mm
W ≥ 25 mm	5 mm

Excess weld metal refers to the metal that lies outside the surface of a mitre fillet or outside a straight line between the toes of the butt/groove weld. This excess metal is sometimes wrongly called reinforcement. This is incorrect because in only very rare cases would it increase the static strength of a joint and in many cases it would reduce the fatigue strength.

Concave contour

The symbol for a concave weld ◡ is only used in special cases, for example, if a welded vessel requires smooth surfaces for ease of cleaning or for surface treatment such as painting. Its location is shown in **Fig. 8.6**.

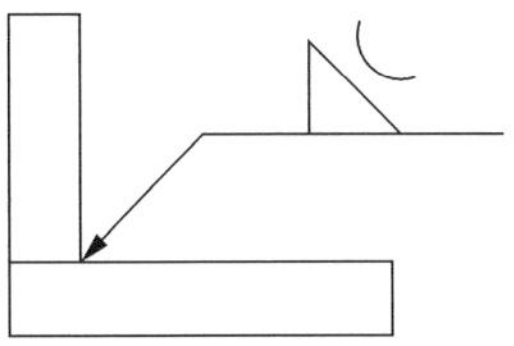

8.6 Location of the concave weld symbol.

A concave weld profile reduces stress concentration at the toes of a fillet weld and thus gives a slight improvement in the fatigue strength.

The ability to obtain a concave weld profile in the as-welded condition depends on the parent metal and the welding process and consumable as well as the expertise of the welder. In mechanised welding processes it is sometimes possible to produce a concave weld profile by using suitable welding parameter settings.

Toes blended smoothly

The ISO standard includes a symbol for weld toes to be blended smoothly. Its location is shown in **Fig. 8.7**. It can be used to inform the welder that the weld toes are to be ground in order to remove any small slag intrusions that exist at the toes of welds made by manual metal arc (MMA) or shielded metal arc (SMAW) welding. The maximum depth of intrusions is usually 0.4 mm (1/64 in) and the depth of grinding should be 1–2 mm (1/32–5/64 in).

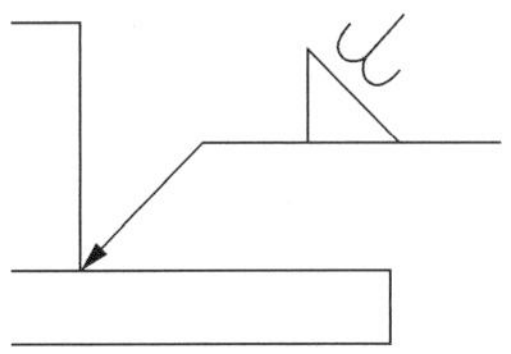

8.7 Location of the symbol for toes blended smoothly.

The purpose of weld toe grinding is to increase the fatigue strength of the welded joint. This is important because slag intrusions can act as initiation sites for fatigue cracks. The process of weld toe grinding for fatigue strength improvement is highly skilled and requires training.

It will be evident from the foregoing descriptions that, if a particular weld profile is desired, it may not be possible to convey all the essential requirements by means of welding symbols. In this case separate, detailed instructions should be given in a Welding Procedure Specification (WPS) or on a note on the drawing.

Other supplementary symbols

The symbols in Fig. 8.1 that do not concern weld contours are a clear instruction to the welder. It is appropriate that these instructions should be included as part of a welding symbol.

The AWS standard includes four symbols that are not used in the ISO system. These are the symbols for a spacer, a back weld or backing weld, melt through and consumable insert. Other supplementary symbols in use, and included in the ISO and AWS standards, are those for peripheral welds, field or site welds and backing strips.

Spacer

The purpose of a spacer ─┤▭├─ is shown in **Fig. 8.8** in which the symbol for a double-V groove weld is modified to indicate the use of a spacer in the joint. The dimensions and material of the spacer are specified in the tail of the reference line or on notes on the drawing. In the figure, a carbon steel spacer measuring 12 × 6 mm (1/2 × 1/4 in) is used.

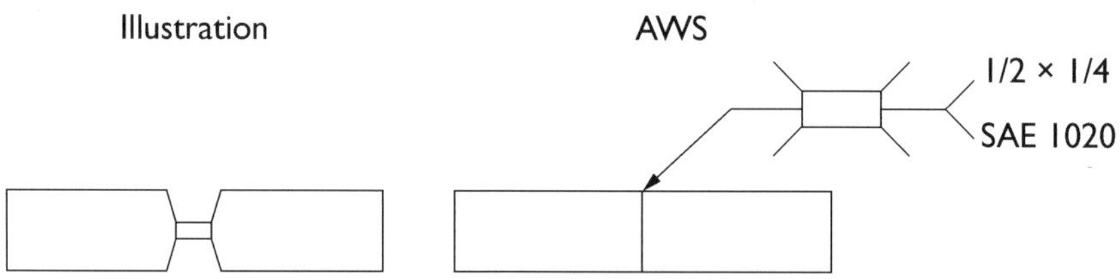

8.8 Spacer (AWS) (dimensions are in inches).

Back weld and backing weld

The symbol ⌒ is used for both a back weld and a backing weld (backing run). A back weld is made on the reverse side of a groove/butt weld after the main weld is completed. A backing weld is made before the main weld is made.

Whether the symbol refers to a back weld or a backing weld or run can be indicated in the AWS system by writing BACK WELD or BACKING WELD in the tail of the reference line as shown in **Fig. 8.9.** This is shown for a single-V groove/butt weld, which also shows an alternative AWS system that uses two reference lines indicating the sequence of operations. In both cases the first operation is indicated by the reference line closer to the arrow.

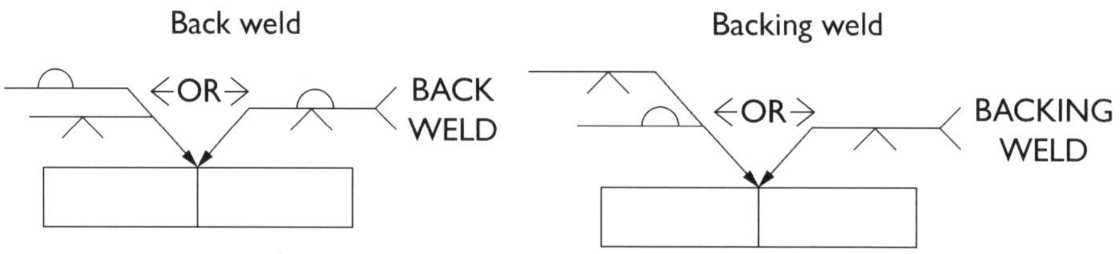

8.9 Back weld and backing weld (backing run) (AWS).

Melt through

The melt through symbol is used when complete penetration is required in welds made from one side, the symbol being placed on the side of the reference line opposite to the main welding symbol. **Figure 8.10** shows a single-V groove weld made from one side with complete joint penetration.

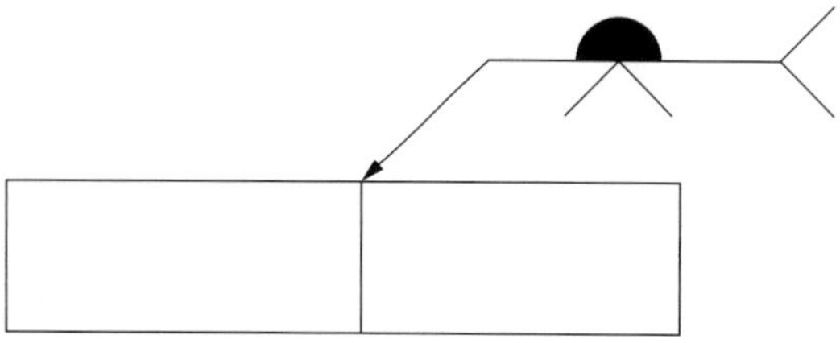

8.10 Melt through (ISO).

The ISO standard requires that all butt welds shall have complete penetration unless there are any contrary indications.

In the AWS standard the melt through symbol is used when complete joint penetration is required plus so-called visible root reinforcement. The height of the root reinforcement may be specified by placing the required dimension to the left of the melt through symbol. Therefore, it is possible to specify the size of the penetration bead, for example 1/16 in or 1/8 in.

The reason for this is difficult to fathom because complete penetration is always visible and cannot be improved on by specifying its size. It would be unfortunate if a product was rejected because, on inspection, penetration was found to be 1/16 in when 1/8 in was specified. Remedial action would be expensive and unnecessary.

If the size of a penetration bead were to be restricted to a certain dimension for clearance purposes this would be an important requirement for which instructions should be given in notes on the drawing.

Consumable insert

Consumable inserts are specified in the AWS system by placing the symbol on the side of the reference line opposite the groove weld symbol, as illustrated in **Fig. 8.11**. The type of insert is written in the tail of the reference symbol or on a note on the drawing.

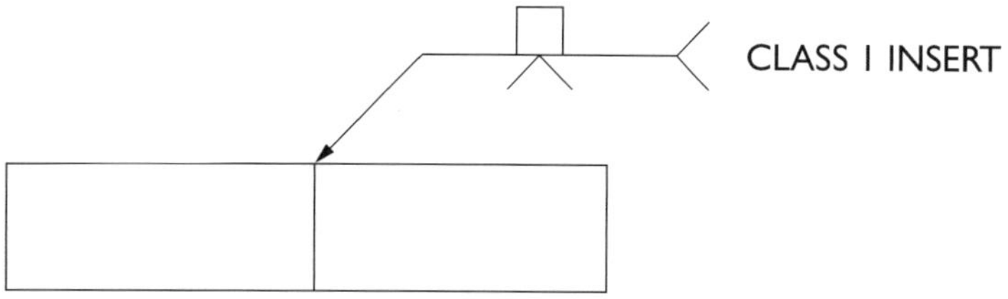

8.11 Consumable insert (AWS).

Peripheral welds (weld all round)

Figure 8.12(a) shows an end plate welded to a rolled steel joist, the symbol indicting a continuous weld round the end of the joist.

Figure 8.12(b) shows the symbol indicating a weld all round a patch plate.

In cases where a continuous weld is required, which involves an abrupt change in direction, for example, round an internal or external corner, the weld all round symbol may be confusing because the extent of welding may not be obvious. This is shown in **Fig. 8.12(c)** where a stiffened beam contains a cope hole at the inside corner of the stiffener. The weld all round symbol would not provide as clear an instruction as the symbol for a double fillet weld in the flat, vertical and overhead welding positions. The multiple arrow system is quite clear and it would be permissible to omit the dashed line in the ISO example.

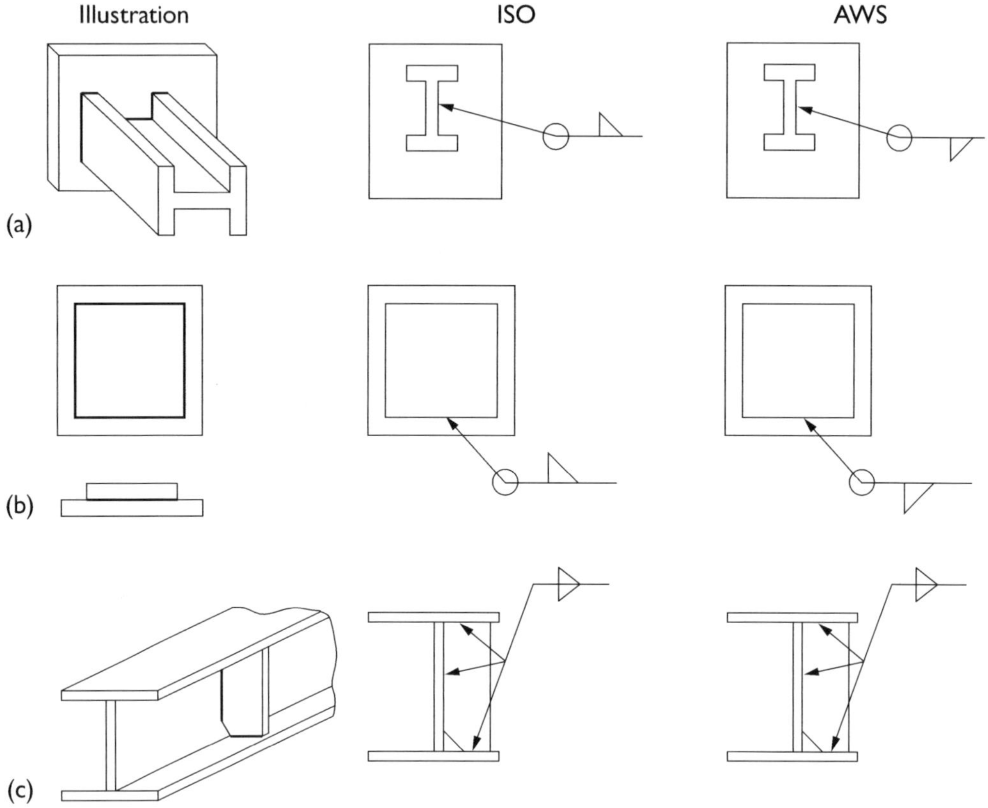

8.12 Peripheral welds for (a) an end plate welded to a rolled steel joist, (b) a patch plate and (c) a stiffened beam.

Field or site weld

Field or site welds, i.e. welds not made in a fabrication shop, are specified by adding the flag symbol to either side of the reference line at the junction with the arrow line, as shown in **Fig. 8.13**. There is no significance in the flag being placed either above or below the reference line or whether it points left or right.

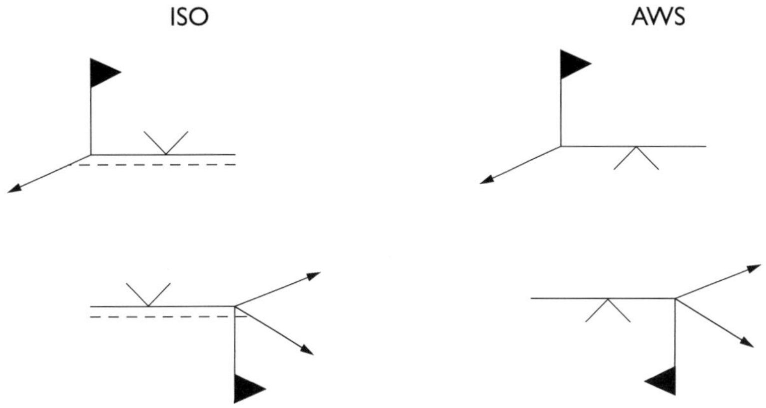

8.13 Field or site weld.

Backing strip or backing

The term backing strip (or backing) should not be confused with back weld or backing weld.

The symbol for a backing strip is the same in the ISO and AWS systems, but the nomenclature is different (see **Fig. 8.14**). A joint with a backing strip is denoted by placing the symbol on the side of the reference line opposite the butt/groove weld symbol and in both standards the letter R indicates that the backing is to be removed after welding.

In the AWS standard the material and the dimensions of the backing strip are specified in the tail of the reference line.

In ISO 2553: 1992, there is no explanation of the meaning of the letter M in the symbol, but it probably originates from previous editions of ISO 2553 or associated standards. The letter M probably refers to the material and dimensions of the backing strip. This information would presumably be included in the Welding Procedure Specification (WPS).

8.14 Single-V weld with removable backing.

9 Dimensions I

Butt/groove welds

In both the ISO and AWS systems, the size of a butt/groove weld can be specified. The AWS standard also specifies edge preparations. The comprehensive description of a joint specified by the welding symbol in the AWS standard should be used with caution unless it is an established part of a welding procedure. Otherwise, the procedure might not be in accordance with the standard practice of a particular fabrication shop. The edge preparation will depend on the welding process, the consumable, the parameters and other details, which may not be known to the design office. In the ISO standard it is stated that, in the absence of any indication to the contrary, butt welds are to have complete penetration (see **Fig. 9.1**).

The AWS standard states that omitting the depth of bevel and weld dimensions from the welding symbol requires complete joint penetration. This is in agreement with the ISO standard. The AWS standard states that the rule applies to single groove welds and also to double groove welds having symmetrical joint geometry (single and double groove welds include V, U, bevel and J welds). The ISO and AWS requirements are, in fact, the same because in both systems asymmetrical double butt/groove welds will need to have the depth of the weld specified, at least on the first side.

Figures 9.1 and **9.2** show fully penetrated single and symmetrical double-V butt/groove welds and their symbolic notation. For double sided welds, the dashed reference line may be omitted from the ISO symbol, which makes it identical to the AWS symbol. This only applies to symmetrical joint preparations where depths of welds are not included in the welding symbol.

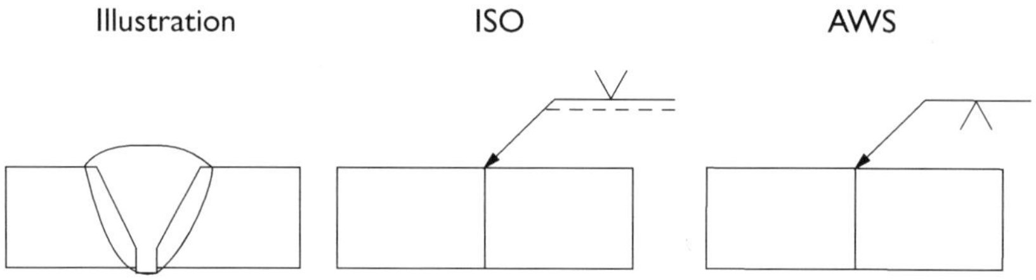

9.1 Single-V butt/groove weld with full penetration.

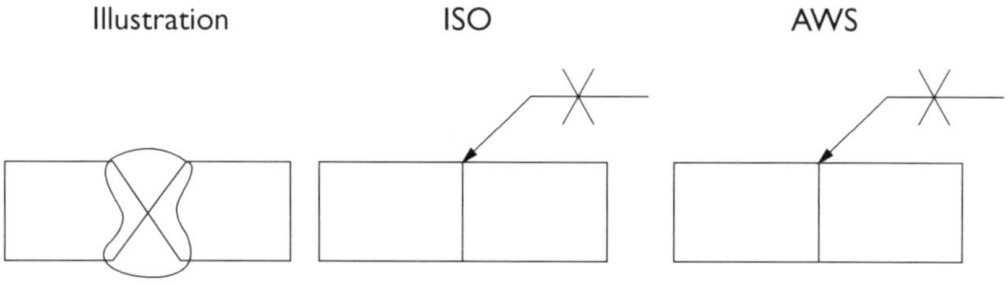

9.2 Double-V butt/groove weld with full penetration.

Partial penetration welds

In both the ISO and AWS systems the size of a butt/groove weld can be specified by a number which is placed to the left of the weld symbol. In the AWS weld symbol, the number is placed in brackets, as in **Fig. 9.3**.

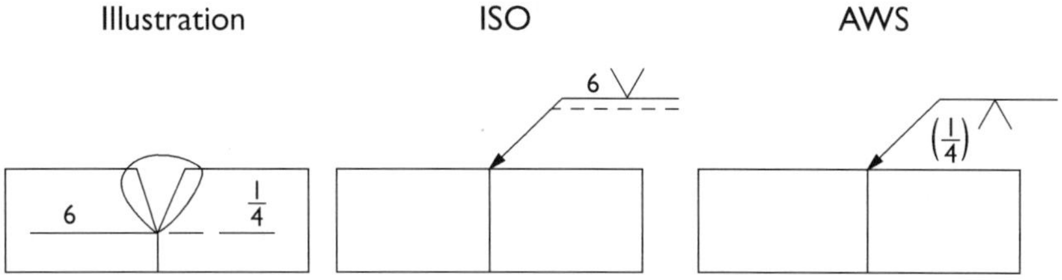

9.3 Partial penetration single-V butt/groove weld. ISO dimensions are in millimetres; AWS dimensions are in inches.

The application of this system to a partial penetration square butt/groove weld is shown in **Fig. 9.4**. In the AWS system, in which edge preparations can be indicated, the size of the gap between the plates is specified by a number between the sides of the weld symbol. The dimensions of the gap need only be included on one side of the reference line.

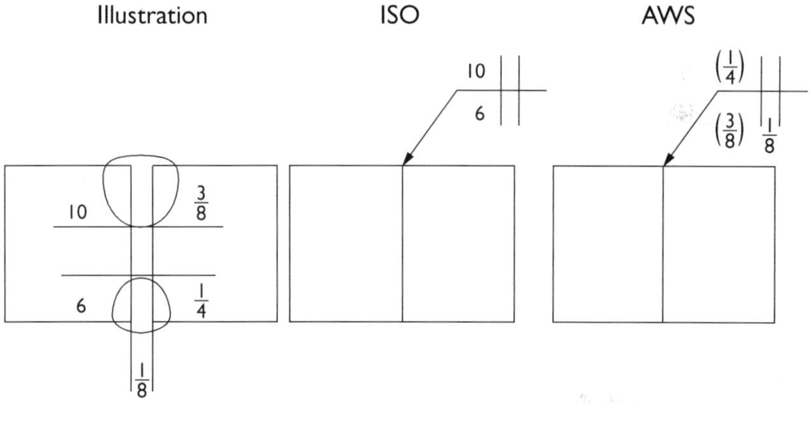

9.4 Partial penetration square butt/groove weld. ISO dimensions are in millimetres; AWS dimensions are in inches.

Groove dimensions

In the AWS system full details of the groove dimensions can be specified, for example, bevel angles, root face, root gap, as well as the sizes of the welds on each side of the joint. **Figure 9.5** shows a 5/8 in thick plate, prepared with 60° included angles, bevel depths of 1/4 in and weld sizes of 3/8 in, giving overlap at the centre of 1/8 in.

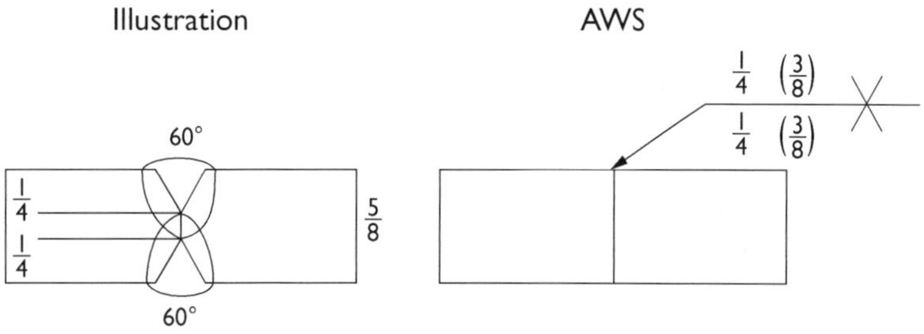

9.5 Depth of bevels and weld sizes (AWS) (dimensions are in inches).

Double butt/groove welds often have the root of the first weld back gouged or ground. This procedure can be specified by the AWS welding symbol. A joint requiring back gouging may be specified by placing the instruction BACK GOUGE in the tail of the reference line.

Figure 9.6 shows a double-V groove weld with details of edge preparation and back gouging. The edge preparation is a double-V with 60° included angles, each side of the joint prepared to depths of 3/8 in. A root face of typically 1/16 in to 1/8 in (not specified by the welding symbol) would be obtained by grinding to produce a root gap of 1/8 in.

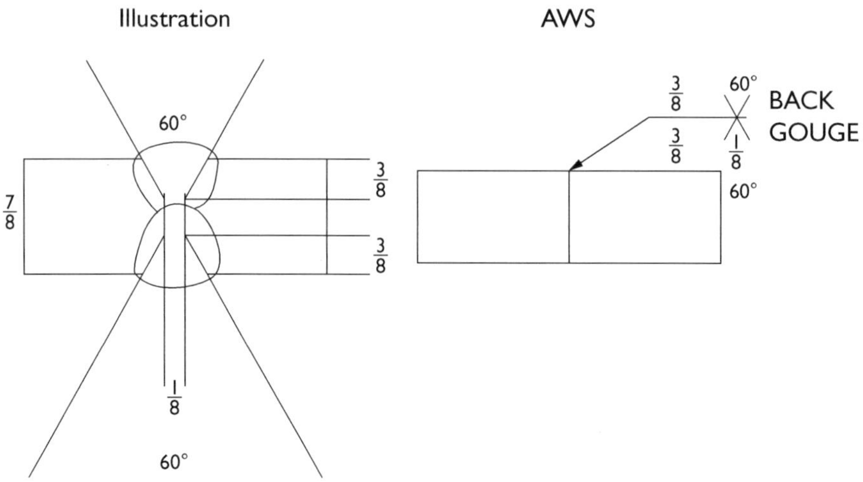

9.6 Double-V groove weld with edge preparation and back gouging instruction (AWS) (dimensions are in inches).

In the example shown in Fig. 9.6, weld sizes are not specified because the depth of the gouge should be sufficient to remove much of the root of the first weld so that full penetration is obtained when the second side is welded. However, it is permissible to specify the size of the weld on the first side to ensure that it is large enough to limit the depth of back gouging required.

The AWS standard describes many examples of combinations of joint edge preparations and weld sizes that can be specified by the welding symbol. In contrast to this, the ISO system limits the information specified by the welding symbol to the type of weld and its size. There is a limit to the extent of a welding procedure that can be contained in a welding symbol and it is advisable to supplement this information with additional notes and drawings.

This is supported by the AWS standard, which recommends that drawing notes may be used to provide instructions for weld details. This information need not be repeated in welding symbols and the information can include details of edge preparations, weld sizes, etc. The ISO standard follows this practice in principle and recommends that, in order to simplify the drawings, specific instructions should be referred to separately or placed close to the welding symbol.

Length of butt/groove welds

When there is no indication of the length of a butt/groove weld, it means that the whole length of the joint is to be welded. Sometimes the weld is shorter than the joint length and may consist of intermittent welds. In this case the specific lengths of welds and their location may be indicated by dimensions on the right of the weld symbol and on the drawing. Hatching may also be used to depict graphically the welds shown in **Fig. 9.7**. This procedure and alternative methods are described in detail in AWS A2.4-98, but not in ISO 2553 which does not include intermittent butt welds. However, the ISO standard does include intermittent fillet welds, described in Section 11, and there is no reason why the principle should not be applied to butt welds symbolised in accordance with the ISO standard.

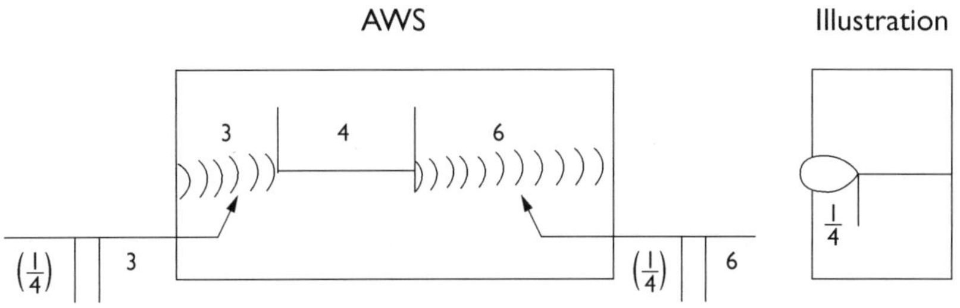

9.7 Intermittent square groove welds with dimensions and hatching (AWS) (dimensions are in inches).

10 Dimensions 2

Fillet welds – transverse

The ISO standard includes two methods to indicate fillet weld sizes: leg length (z) and throat thickness (a) (see **Fig. 10.1**).

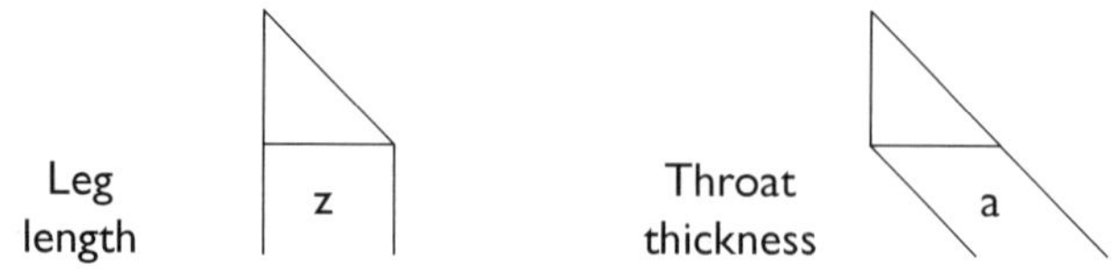

10.1 Fillet weld size (ISO).

In the AWS standard the size of a fillet weld always refers to the leg length, the dimension of which is placed to the left of the weld symbol. In contrast to the symbol for a groove weld, the fillet weld size is not placed in brackets.

In the ISO system, to indicate the size of a fillet weld, the dimension is placed to the left of the symbol, preceded by the letter z or a, depending on whether the leg length or throat thickness is to be specified (see **Fig. 10.2**). It is generally the leg length that is specified.

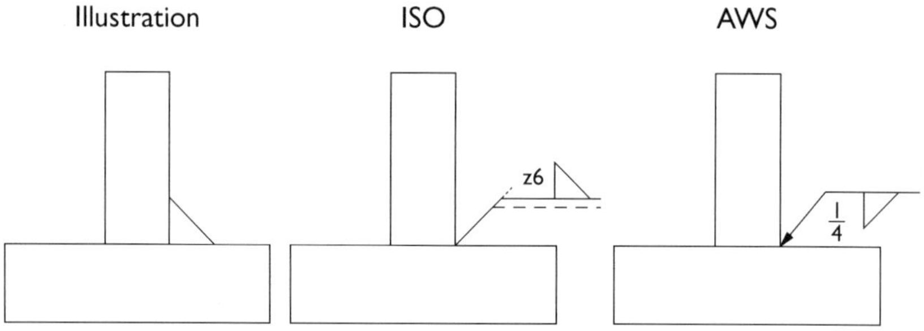

10.2 Fillet weld with 6 mm (1/4 in) leg length.

The terms.describing fillet weld dimensions, actual throat thickness and design throat thickness, are shown in **Fig. 10.3** and are defined in the section Terms and definitions.

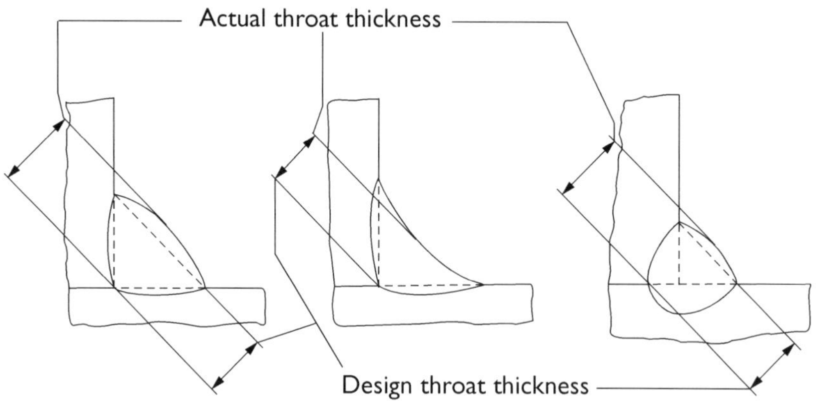

10.3 Actual throat thickness and design throat thickness of fillet welds.

Deep penetration welds

The ISO standard can specify deep penetration welds, the effective weld throat being indicated by the letter s placed in front of the throat thickness dimension, as shown in **Fig. 10.4**. This is followed by the nominal throat thickness preceded by the letter a. The reason for including the nominal throat dimension is not clear because any calculations of weld strength would be based on the 's' dimension which is the actual throat thickness.

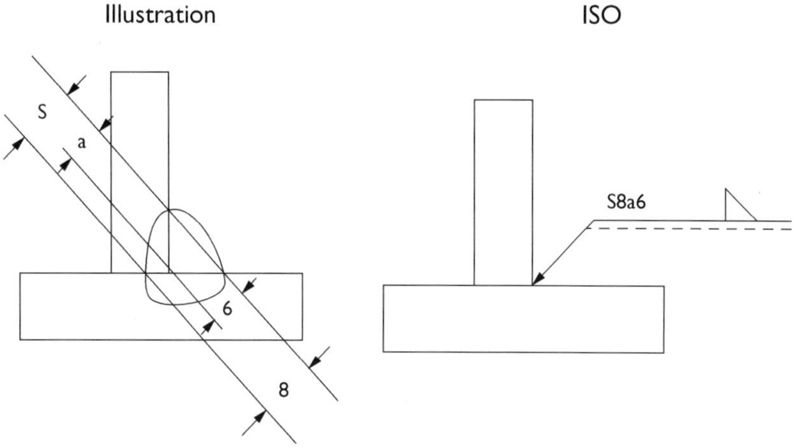

10.4 Deep penetration weld with ISO weld symbol (dimensions are in millimetres).

Deep penetration of a weld can be achieved by high-current submerged arc welding (SAW) or gas shielded metal arc welding (GMAW) and by certain types of electrodes. The depth of penetration must be predetermined by welding procedure testing before the dimension is used for design purposes and is included in the welding symbol.

In the AWS standard deep penetration fillet welds are not mentioned as such and it is not clear how such a weld would be symbolised. The dimensions of the weld would have to be specified on a drawing note and, as mentioned above, procedure testing would be necessary before the weld size was used in design calculations.

Double fillet welds

In a double fillet weld the dimensions are repeated even if they are identical (see **Fig. 10.5**).

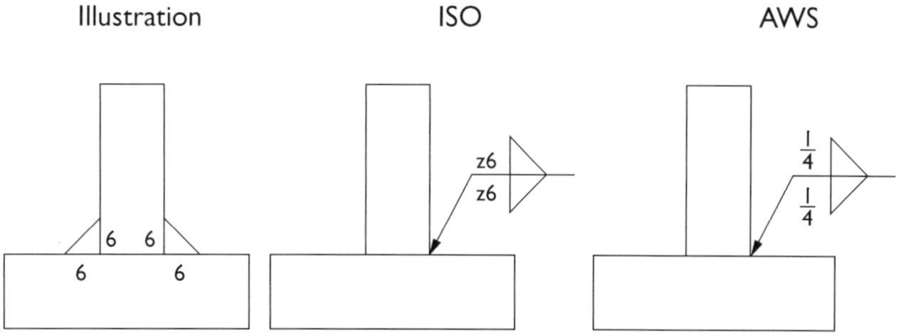

10.5 Double fillet weld. ISO dimensions are in millimetres; AWS dimensions are in inches.

Unequal leg length fillet welds

If it is not obvious which is the short or long leg, identification can be indicated by instructions in the tail of the reference line. The ISO system uses a closed tail with a reference letter which can refer to a specific instruction such as a note on the drawing (see **Fig. 10.6**).

The direct instruction or an unambiguous note on the drawing is preferable because the AWS standard places the smaller dimension of an unequal leg length fillet weld first whereas the ISO standard does not refer to unequal leg length fillet welds at all.

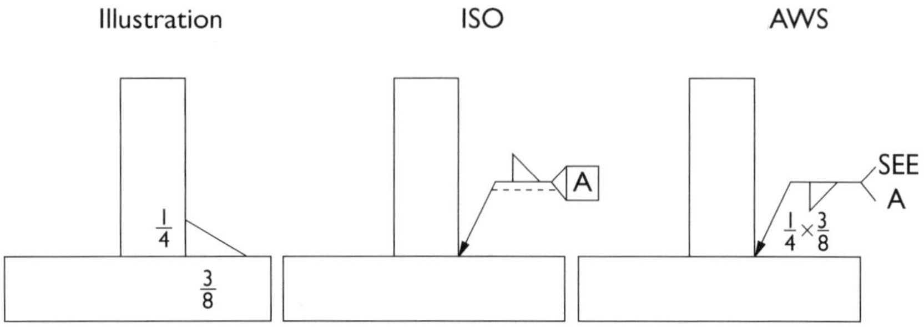

10.6 Weld symbols for an unequal leg length fillet weld. AWS dimensions are in inches.

The AWS system uses a similar device or can include a direct instruction (see **Fig. 10.7**).

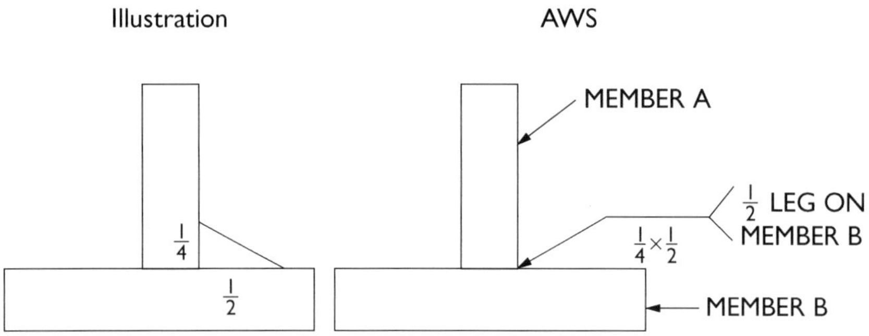

10.7 Detailed instruction with AWS symbol (dimensions are in inches).

11 Dimensions 3

Fillet welds – longitudinal

Dimensions measured along the length of the weld are shown to the right of the weld symbol in both the ISO and AWS standards and, in the absence of any dimensions, the weld is continuous along the whole length of the joint. If the weld is not continuous and consists of intermittent fillet welds, the weld lengths and the gaps between them are indicated as shown in **Fig. 11.1**. This shows a perspective and plan view as well as a cross-section, the latter indicating the leg length and throat thickness of the weld in the example illustrated.

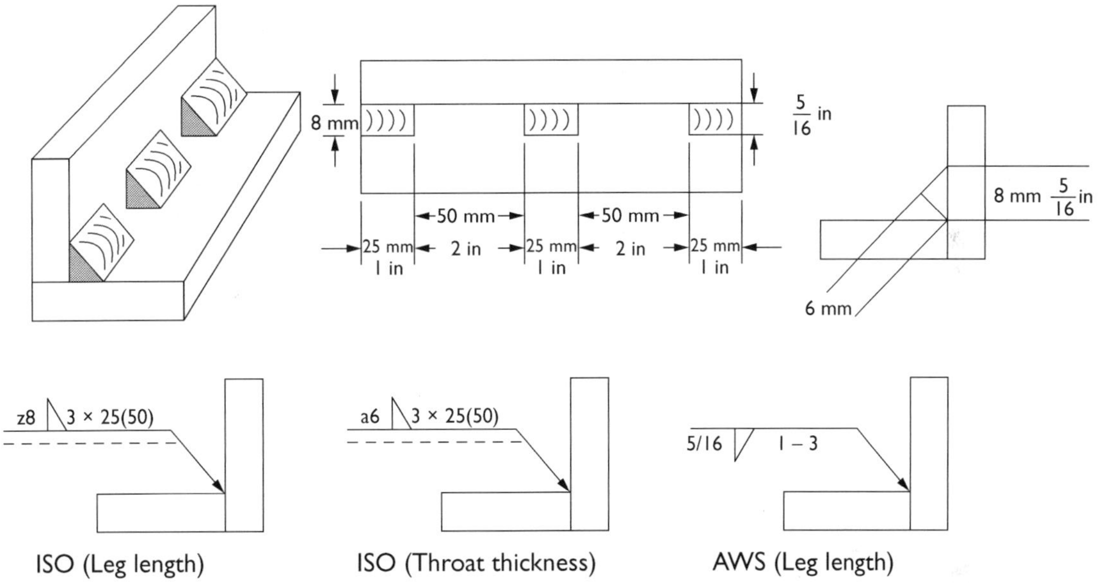

11.1 Dimensions of single fillet welds. ISO dimensions are in millimetres; AWS dimensions are in inches.

The symbolic representation of weld sizes and lengths, and the gaps between them, is shown in the lower three cross-sections.

The nomenclature placed adjacent to the reference lines is shown in **Table 11.1**.

Table 11.1 Nomenclature placed adjacent to the reference lines of intermittent fillet welds in both the ISO and AWS systems

ISO (mm)	AWS (in)
z8 – leg length	5/16 – leg length
25 – length of each weld	1 – length of each weld
(50) – distance between the ends of adjacent welds	3 – pitch (centre to centre spacing)
3 – number of separate welds	

Intermittent fillet welds are used in situations where it is necessary to use a weld, the size of which would give a higher load-bearing capacity than required if used for the whole length of a joint, with consequent extra cost of weld metal.

The use of a smaller weld might not be possible because of limitations of a particular welding process or because of poor fit-up of the joint. Small welds of some alloy compositions may crack under conditions of high restraint. Low heat input associated with such welds can increase the risk of the heat affected zone (HAZ) cracking in steel fabrications.

These factors confirm the necessity for cooperation between the design office and the fabrication shop before the sizes of the welds are included on a drawing.

Figure 11.2 shows the symbolic representation of intermittent double fillet welds. It should be noted that the ISO symbol omits the dashed reference line.

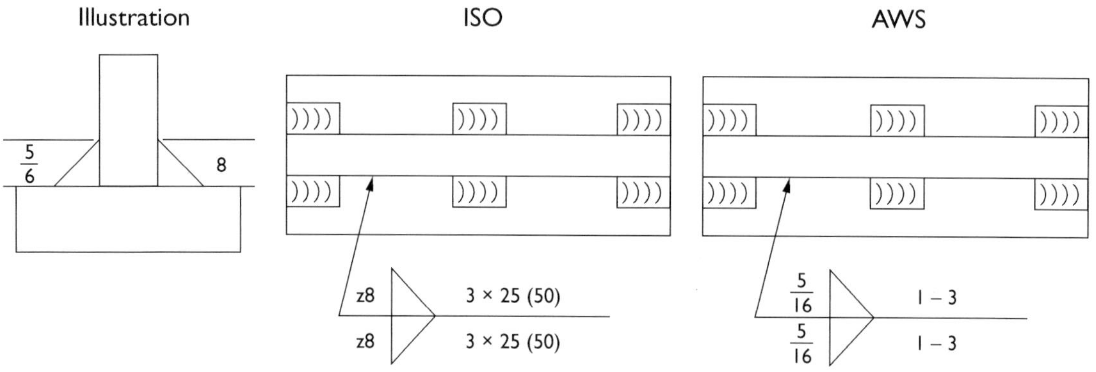

11.2 Dimensions of intermittent double fillet welds (weld lengths and gaps as in Fig. 11.1).

Figure 11.3 shows staggered intermittent fillet welds which are indicated on the ISO system by an elongated 'Z'. In the AWS standard, the staggered mode is indicated by the displacement of the fillet weld symbols on the reference line.

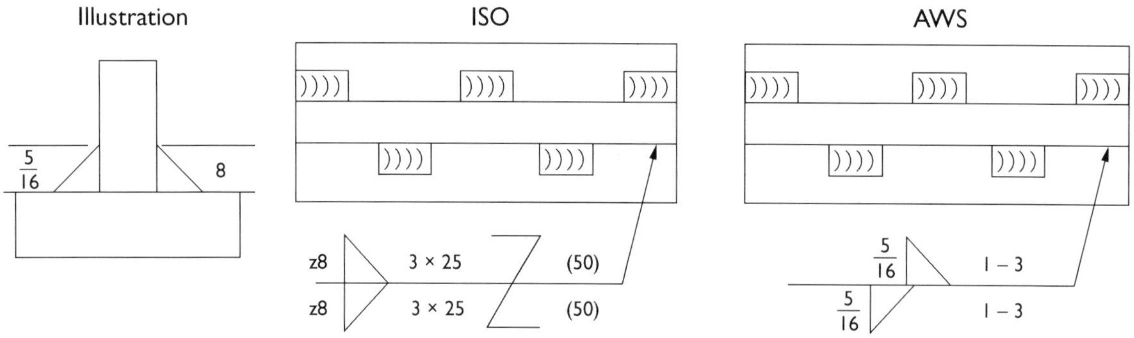

11.3 Dimensions of staggered intermittent fillet welds (weld lengths and gaps as in Fig. 11.1).

As an aid to location, intermittent fillet welds may be graphically located by hatching as shown in **Fig. 11.4**. This method is described in the AWS standard and illustrated in the ISO standard. It is a useful aid to the interpretation of welding symbols.

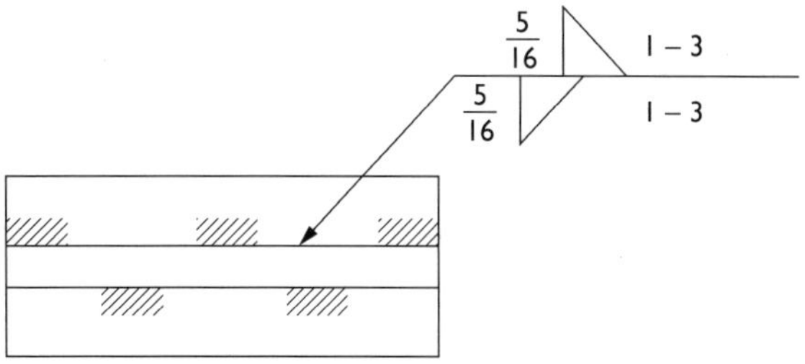

11.4 Hatching showing location of fillet welds (AWS) (dimensions are in inches).

12 Spot and seam welds

Resistance spot welds

Resistance spot welding requires access to both sides of the joint. The weld symbol is centred on the reference line (see **Fig. 12.1**).

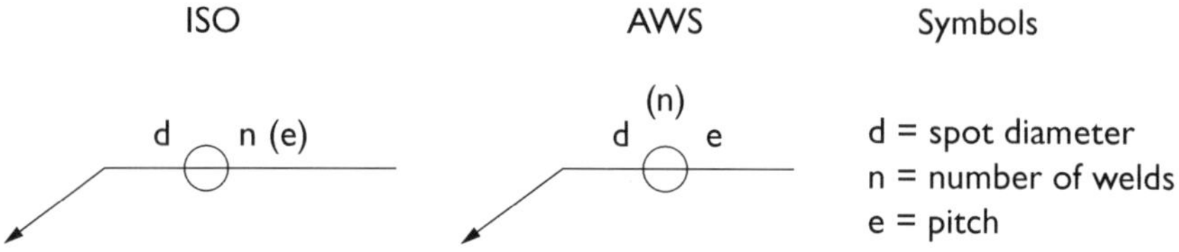

12.1 Symbolisation of a resistance spot weld.

Other details of the welding symbol shown in Fig. 12.1 are compared in **Table 12.1**.

Table 12.1 Symbols for dimensions and spacing of resistance spot welds

Details	ISO	AWS
Spot diameter at interface	d	d
Number of welds	n	(n)
Pitch	(e)	e

An example of the application of the welding symbol is shown in **Fig. 12.2**. Under the heading 'Strength', AWS A2.4-98 specifies that the strength of a spot weld shall be indicated by a number quoting pounds or newtons placed in front of the weld symbol. The strength of a spot weld is related to its size and it is questionable whether mechanical properties should form part of a welding symbol. The purpose of this option and for whom the information is intended are not clear.

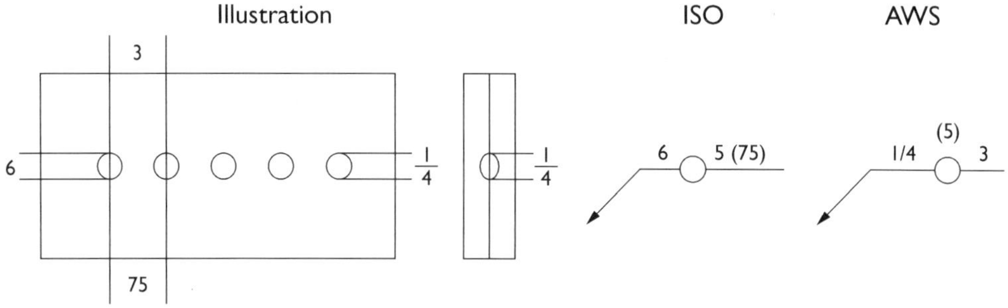

12.2 Examples of resistance spot weld symbols. ISO dimensions are in millimetres; AWS dimensions are in inches.

Arc spot welds

Arc spot welding is carried out from one side of the joint so the weld symbol is placed on one side of the reference line as shown in **Fig. 12.3**. This figure also shows the different ways of indicating the number of welds in the ISO and AWS standards.

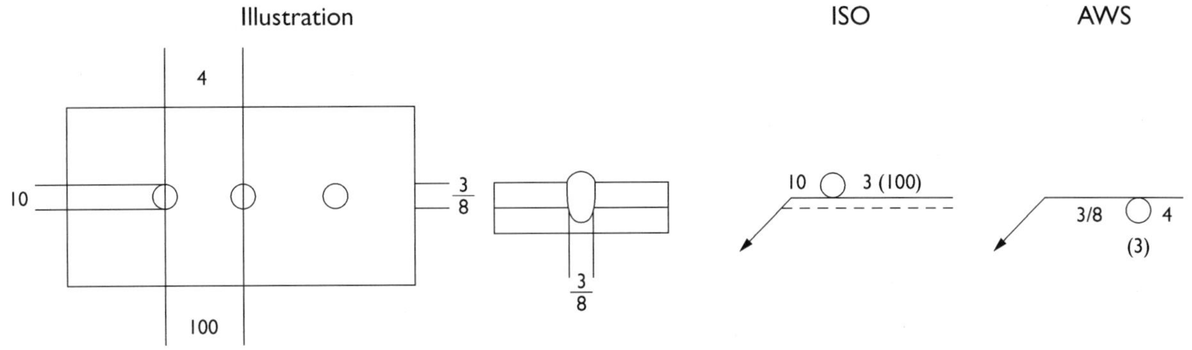

12.3 Examples of arc spot weld symbols. ISO dimensions are in millimetres; AWS dimensions are in inches.

The nomenclature for both resistance and arc spot welds in the ISO and AWS systems is similar, apart from the placement of the brackets and the figure for the number of welds.

Projection welds

In both the ISO and AWS standards the weld symbols for spot and projection welds are identical, but have additional indications. In the ISO system the letter P is placed in front of the weld size and in the AWS standard the letters PW are placed in the tail of the reference line. A reference to the size of the projection by a separate sketch is indicated in the tail in the AWS system, as shown in **Fig. 12.4**. This figure also shows that the weld symbol in the AWS system is placed above or below the reference line, depending on which member is to be embossed.

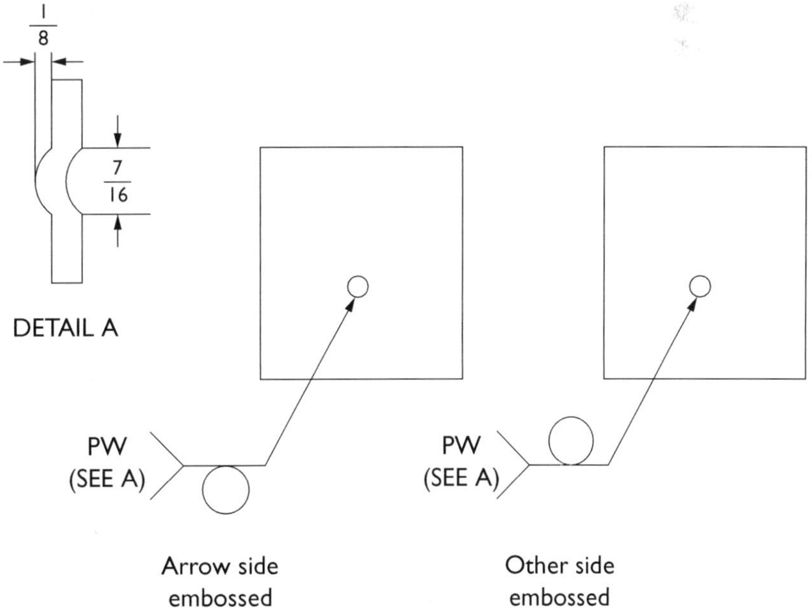

12.4 Symbolisation of projection weld (AWS) (dimensions are in inches).

Seam welds

The notation for the sizes and spacing of intermittent seam welds is shown in **Fig. 12.5**. The only difference in the meaning of the letters is that in the AWS system the letter e indicates the pitch of the welds, i.e. centre to centre, whereas in the ISO standard the bracketed letter (e) refers to the distance between welds.

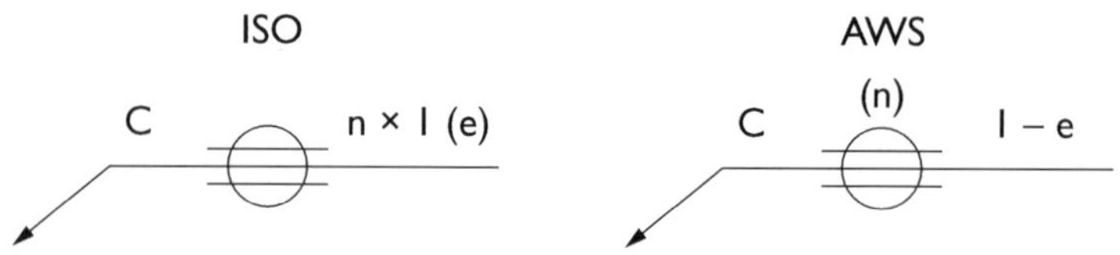

Width of weld at interface	ISO	AWS
Length of each weld	C	C
Distance between welds	l	l
Pitch of welds	(e)	-
Number of welds	-	e
	n	(n)

12.5 Symbolisation of resistance seam welds.

The seam welding symbol in the ISO system refers to resistance welding, whereas in the AWS standard, the symbol includes other processes such as arc or electron beam welding. Therefore, in the ISO system the symbol is always placed with its centre on the reference line. In the AWS system it may be placed above or below the line, depending on whether the weld is made from the arrow side or the other side.

AWS A2.4-98 gives comprehensive descriptions of the symbolisation of seam and other resistance welding processes. This standard should be consulted for authoritative guidance.

13 Stud welds

The stud welding processes, which are used to attach studs to a component in order to locate or secure other parts, are generally applied to studs ranging from 1.6 mm (1/16 in) to 19 mm (3/4 in) in diameter. Studs larger than this can be attached by manual welding using covered electrodes to deposit a fillet weld. These welds are indicated on a drawing by the fillet weld symbol and the weld all round symbols shown for the ISO example in **Fig. 13.1**.

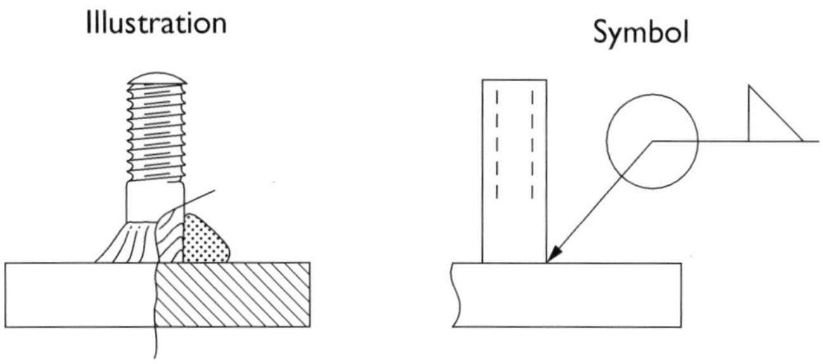

13.1 Fillet weld and peripheral weld symbols for stud welding attached by MMA (SMAW) process (ISO).

Stud welds are not specifically mentioned in ISO 2553, but the standard includes a weld symbol for a surface joint as shown in **Fig. 13.2**. This appears to show a stud weld. The application of this symbol on a drawing is not illustrated in ISO 2553 but, presumably, the two horizontal lines are placed above and parallel to the reference line.

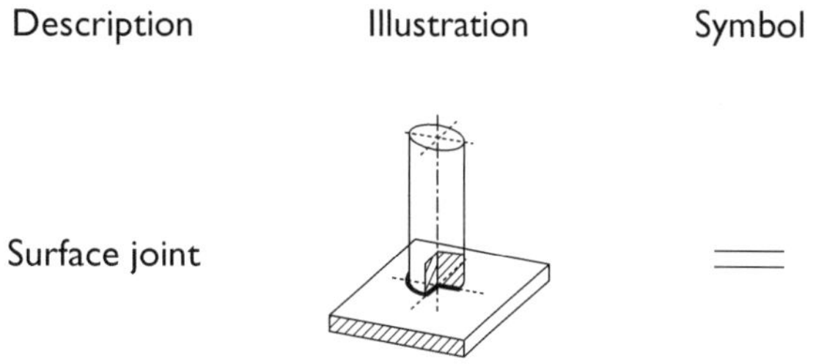

13.2 Weld symbol for a surface joint (ISO).

The AWS symbol for a stud weld consists of a circle with a cross in it. The standard specifies that the arrow of the welding symbol shall point clearly to the surface to which the stud is to be welded. Therefore the symbol is always placed below the reference line to indicate that the weld is on the arrow side. The AWS standard illustrates the use of multiple arrow lines pointing to different rows of studs with their locations indicated on the drawing. The diameter of the stud is placed to the left of the weld symbol and the spacing of studs to the right. The number of studs is placed in brackets below the symbol as shown in **Fig. 13.3**.

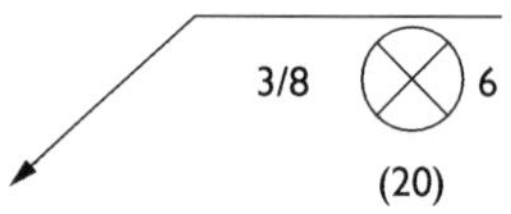

13.3 Symbol for 20 studs, 3/8 inch in diameter with 6 inch spacing (AWS).

As well as, or in place of, the weld symbol, a stud weld can be indicated by code numbers or letters, described in Section 15. The ISO code for stud welding is 781 and the letters used by AWS are SW, these indications being placed in the tails of the reference lines (see **Fig. 13.4**). It should be noted that the numbers and letters in the tail do not provide complete information about the welding process, which may be capacitor discharge, arc stud, friction welding, etc. This information should be given in a note on the drawing or in a Welding Procedure Specification (WPS).

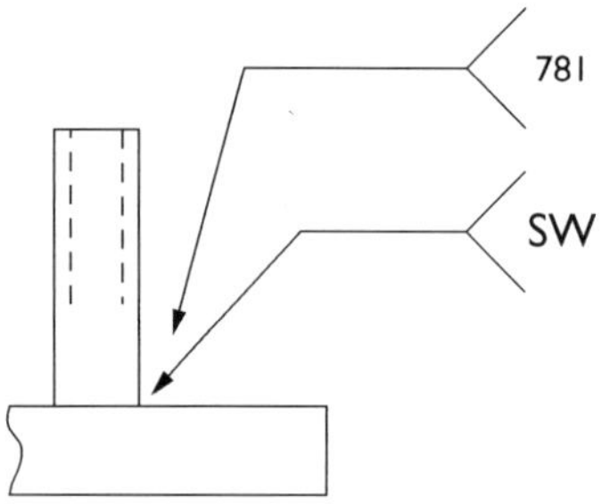

13.4 ISO numbers and AWS letters for stud welds.

14 Surfacing

The surfacing weld symbol can be used to indicate hard facing with wear-resistant materials, cladding for corrosion resistance, or the build-up of surfaces to required dimensions.

The surfacing symbol should point clearly to the surface on which the surfacing weld is to be deposited. The thickness of the weld is specified by placing the dimension to the left of the weld symbol, the area to be surfaced is indicated on the drawing (see **Fig. 14.1**).

Multiple layers

Multiple layers of surfacing welds may be specified by multiple reference lines, with the thickness of each layer specified by placing its dimension to the left of the weld symbol as shown in **Fig. 14.2**.

Figures 14.1 and 14.2 show AWS welding symbols which are described in detail in AWS A2.4-98, but the same method could be used with ISO welding symbols.

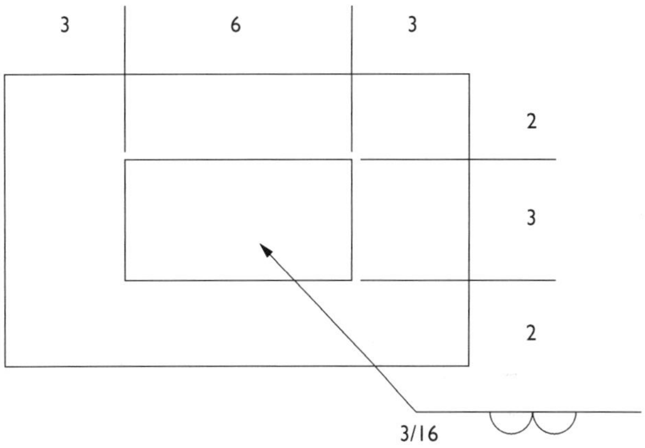

14.1 Symbol for surfacing part of a surface (AWS) (dimensions are in inches).

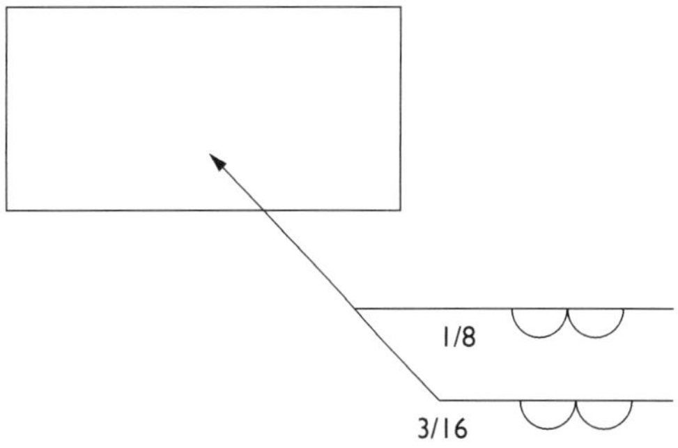

14.2 Symbol for surfacing the whole surface with two layers (AWS) (dimensions are in inches).

15 Process identification

In the ISO system the welding process to be used can be symbolised by a number written within the fork or tail at the end of the reference line, whereas in the AWS system letters are used instead of numbers. An example is given in **Fig. 15.1** where plasma arc welding is represented by the number 15 and the letters PAW, respectively, in the two systems.

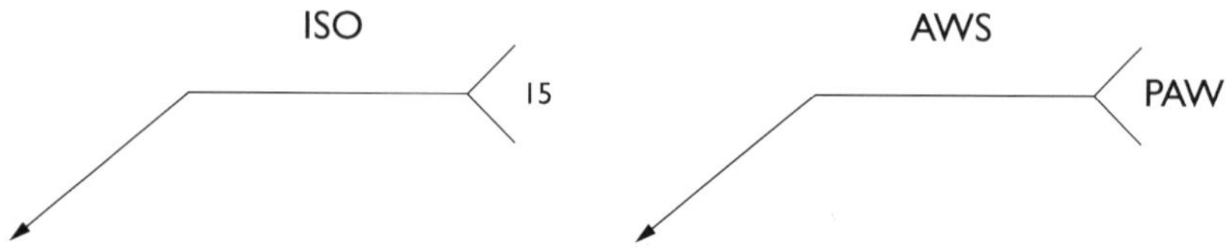

15.1 Process identification.

Table 15.1 gives a comparative list of the ISO and AWS designations for commonly used welding processes. A complete list of the numbers for different welding processes is given in ISO 4063 and for the letter symbols in ANSI/AWS A2.4-98.

Table 15.1 also includes suffixes which can be inserted following the AWS welding process designation, eg. robotic gas tungsten arc welding would be indicated by GTAW–AU.

Apart from process identification, other information, such as codes or standards, may be specified by placing a reference in the tail of the symbol. It will be appreciated that the provision of such comprehensive information by means of welding symbols could lead to confusion and, in most cases, the information should be detailed on a separate part of the drawing and in a Welding Procedure Specification (WPS).

When information is not required in the tail of the reference line, it is standard practice to omit the tail in the AWS system. It appears that this practice is also followed in ISO 2553.

Table 15.1 Designations of welding processes

ISO 4063 : 1992		**ANSI/AWS A2.4–98**	
1	Arc welding	AW	Arc welding
111	Metal arc welding with covered electrode (manual metal arc welding)	SMAW	Shielded metal arc welding
114/136	Flux cored metal arc welding	FCAW	Flux cored arc welding
12	Submerged arc welding	SAW	Submerged arc welding
13	Gas shielded metal arc welding		
131	MIG welding	GMAW	Gas metal arc welding
135	MAG welding (non-inert gas)		
141	TIG welding	GTAW	Gas tungsten arc welding
15	Plasma arc welding	PAW	Plasma arc welding
2	Resistance welding	RW	Resistance welding
21	Spot welding	RSW	Resistance spot welding
22	Seam welding	RSEW	Resistance seam welding
23	Projection welding	RPW	Projection welding
24	Flash welding	FW	Flash welding
3	Gas welding	OFW	Oxyfuel gas welding
311	Oxy-acetylene welding	OAW	Oxyacetylene welding
42	Friction welding	FRW	Friction welding
43	Forge welding	FOW	Forge welding
71	Thermit welding	TW	Thermit welding
72	Electroslag welding	ESW	Electroslag welding
781	Arc stud welding	SW	Stud arc welding
91	Brazing	B	Brazing
94	Soldering	S	Soldering
97	Braze welding		
		Suffixes	
		MA	manual
		SA	semi-automatic
		AU	robotic
		ME	machine

16 Non-destructive testing symbols – AWS

ANSI/AWS A2.4-98 includes a comprehensive system to indicate the requirements for non-destructive testing (NDT) on a drawing, but in ISO 2553: 1992 there is no reference to NDT.

The AWS letter designations for NDT are listed in **Table 16.1**.

Table 16.1 The AWS letter designations for non-destructive testing (NDT)

Type of test	Symbol
Acoustic emission	AET
Electromagnetic	ET
Leak	LT
Magnetic particle	MT
Neutron radiographic	NRT
Penetrant	PT
Proof	PRT
Radiographic	RT
Ultrasonic	UT
Visual	VT

A note of caution is necessary regarding the visual testing symbol VT because many drawings, probably the majority, will be issued without symbols for NDT. This should not be taken to imply that visual inspection is not required. It is essential and should be obvious that all welded components should be visually inspected as a matter of routine.

An application of symbols for NDT is shown in **Fig. 16.1** in which the symbols are placed on a reference line, as for weld symbols, with an arrow line pointing to the welded joint.

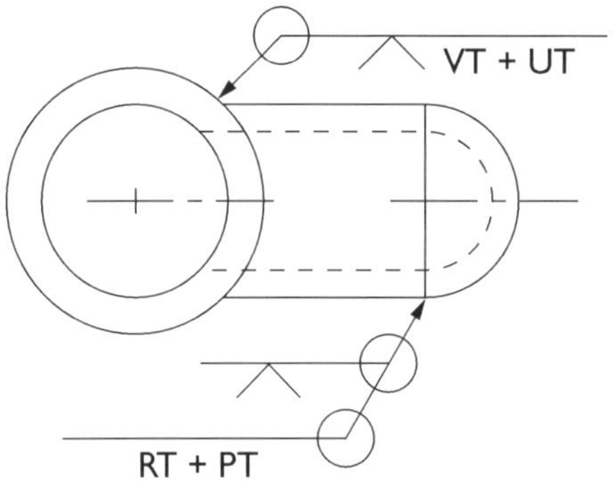

16.1 Welding and non-destructive testing (NDT) symbols for a branch to cylinder weld and a cap to branch weld (AWS).

It should be noted that, in contrast to the arrow for welding symbols, the arrow for NDT always points to the weld to be tested and never refers to a weld on the other side of the joint. If a double fillet weld is to be tested, it would be indicated by two arrows, one on each side of the joint. Figure 16.1 shows that NDT and weld symbols may be combined on the same reference line or, alternatively, two reference lines may be used.

The direction of radiation for radiographic testing may be indicated (see **Fig. 16.2**). The field test symbol (**Fig. 16.3**) and the test all-round symbol (**Fig. 16.4**) are similar to the corresponding weld symbols.

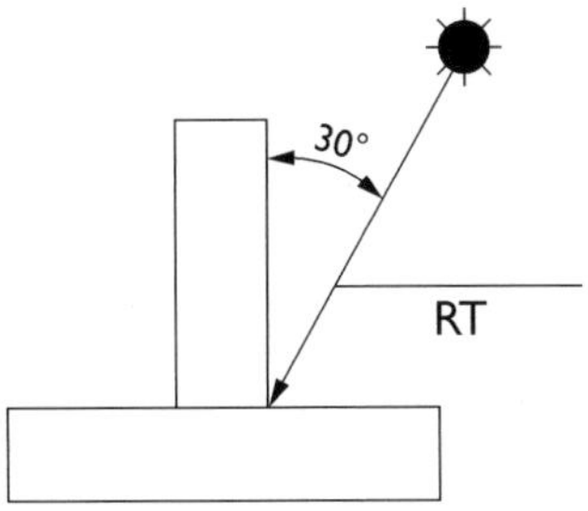

16.2 Symbols for direction of radiographic testing (AWS).

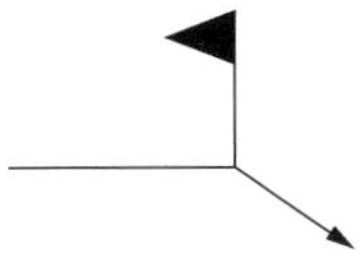

16.3 Field test symbol (AWS).

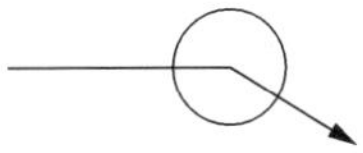

16.4 Test all-round symbol (AWS).

For the clarification of complex drawings the symbols for non-destructive testing can be included on a separate drawing from that showing welding symbols.

An NDT procedure sheet would normally be needed to supplement the welding symbol, especially if inspection is required during welding to confirm freedom from cracking. This could involve dye penetrant or magnetic particle inspection between weld passes, a procedure that the welding symbol alone could not indicate.

Visual inspection should nearly always be carried out during the welding operation to detect any flaws as well as distortion because it is easier to rectify at an intermediate stage than after welding is completed.

17 Exercises

On the following pages the reader is set a number of exercises in both the conversion of information to symbols and the interpretation of drawings in which the weld information is conveyed in symbolic form. The exercises demonstrate the need for cooperation between the designer and fabricator to ensure that the welding procedures are feasible before welding symbols are added to a drawing.

The solutions to the problems begin on page 48. The figures have not been drawn to scale.

Exercise 1: Flange ended pipe

For the fabrication illustrated in **Fig. 17.1**, prepare a drawing in which the welds are described by symbols in accordance with both AWS and ISO representation.

The solution is given on page 48.

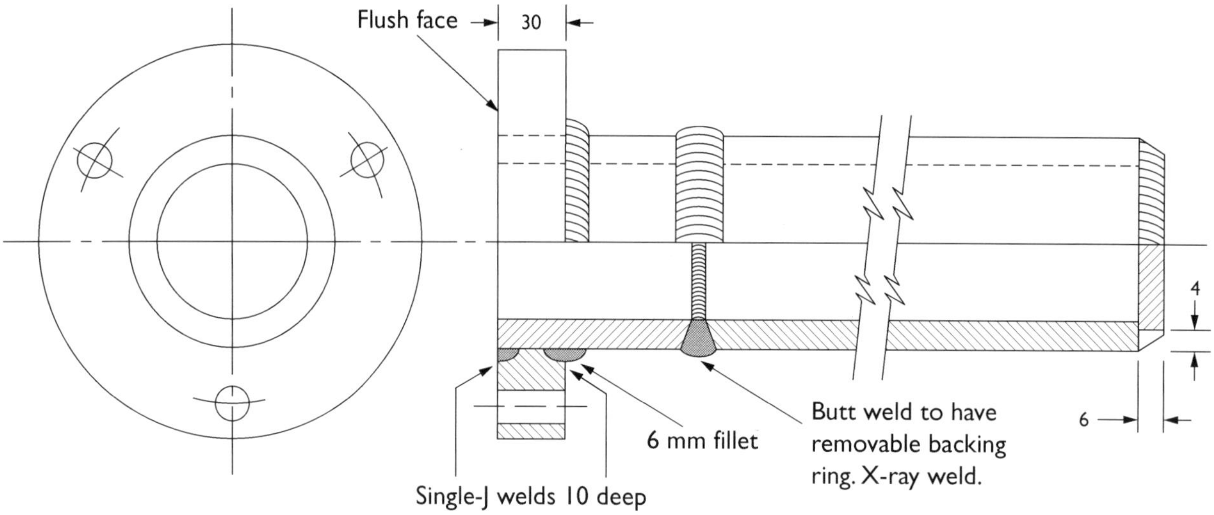

17.1 Exercise 1: Flange ended pipe (dimensions are in millimetres).

Exercise 2: Vessel

Figure 17.2 shows a fabrication with welds to be made, indicated by symbols. The letters in circles are shown solely for ease of reference in the solution.

Problem 1

Describe the welds to be made, using sketches where necessary.

The solution to this problem is given on pages 49 and 50.

Problem 2

Sketch the symbols complying with ISO 2553: 1992, which would be needed to indicate the same welds.

The solution to this problem is given on pages 51 and 52.

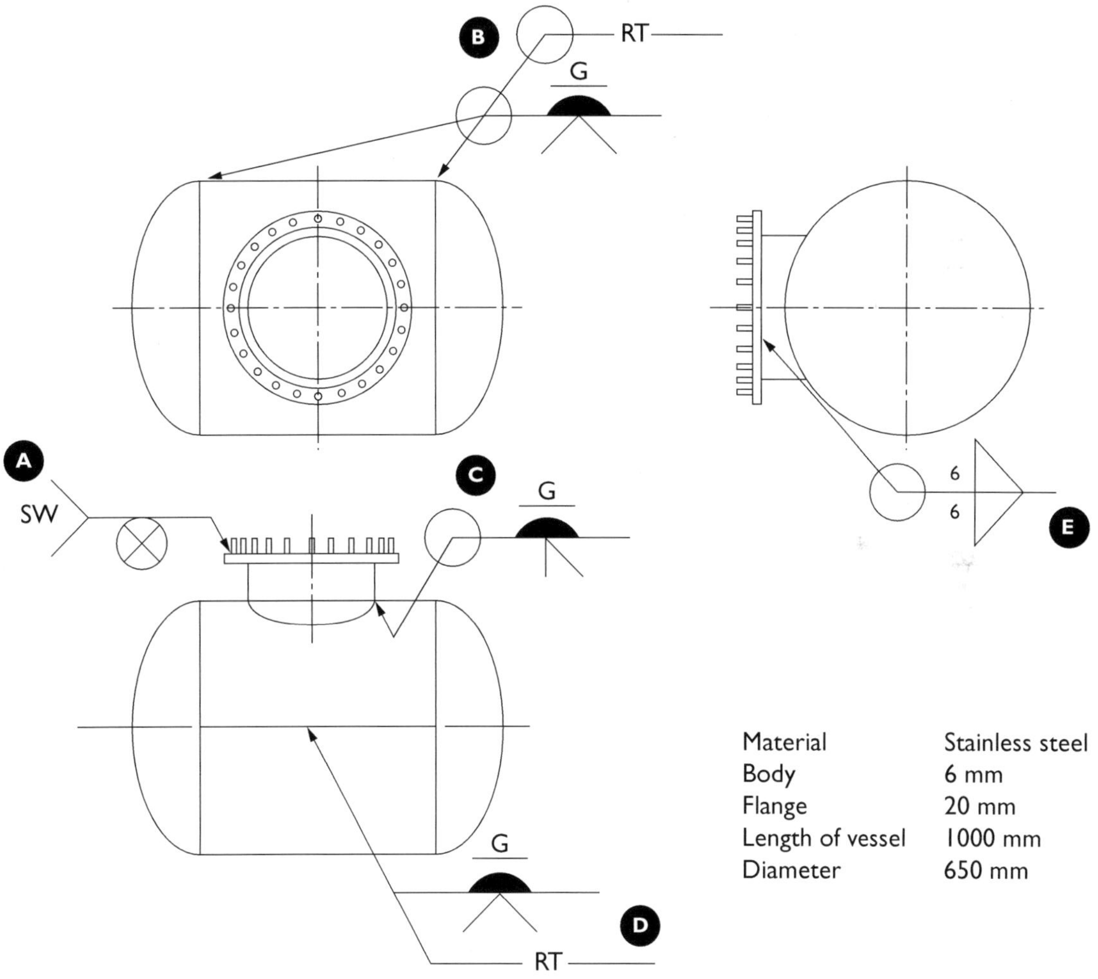

17.2 Exercise 2: Vessel (dimensions are in millimetres).

Exercise 3: Tank

Figure 17.3 shows a tank to be assembled by welding. The letters in the circles are shown only to facilitate reference to the individual joints in the solution.

Problem

Select welding processes and types and indicate by symbols to ISO 2553: 1992.

There are many different solutions to this problem depending on the number of tanks to be manufactured. Consider this tank to be a prototype and that all welding and brazing processes are available.

A possible solution is given on pages 53 and 54.

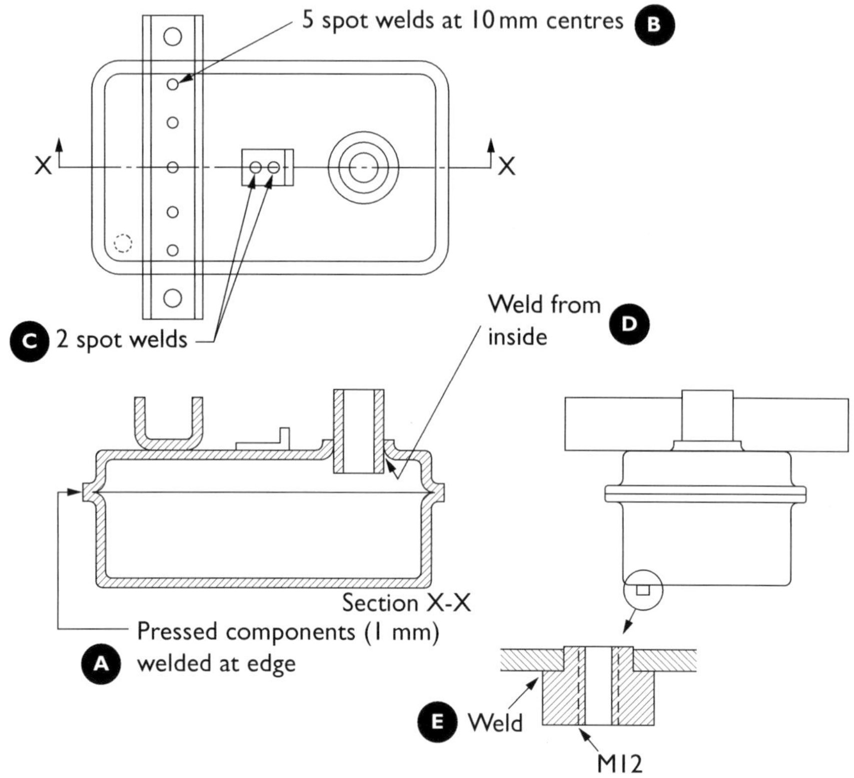

17.3 Exercise 3: Tank.

Exercise 4: Beam

Figure 17.4 shows a sketch, not to scale, of an I-beam which is to be fabricated by manual metal arc (shielded metal arc) welding, carried out both in the shop and on site.

Problem

Make a sketch indicating the welding details and using symbols to ISO 2553: 1992.

The solution is given on page 55.

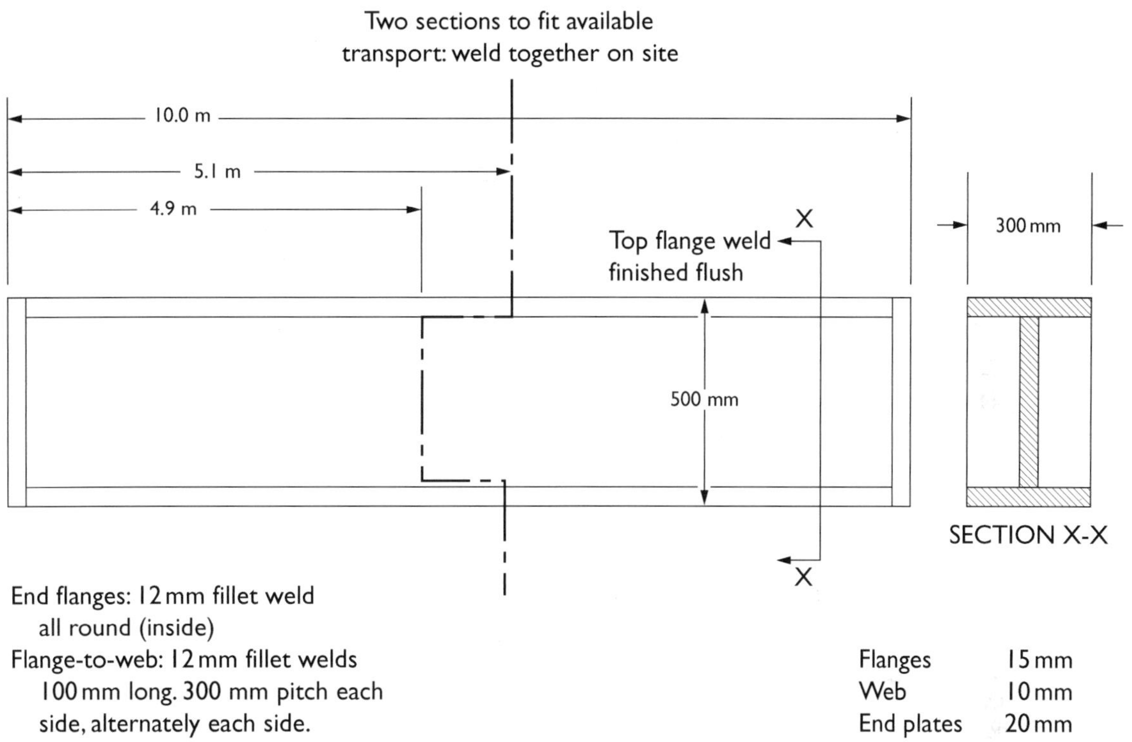

17.4 Exercise 4: Beam.

Exercise 1 solution

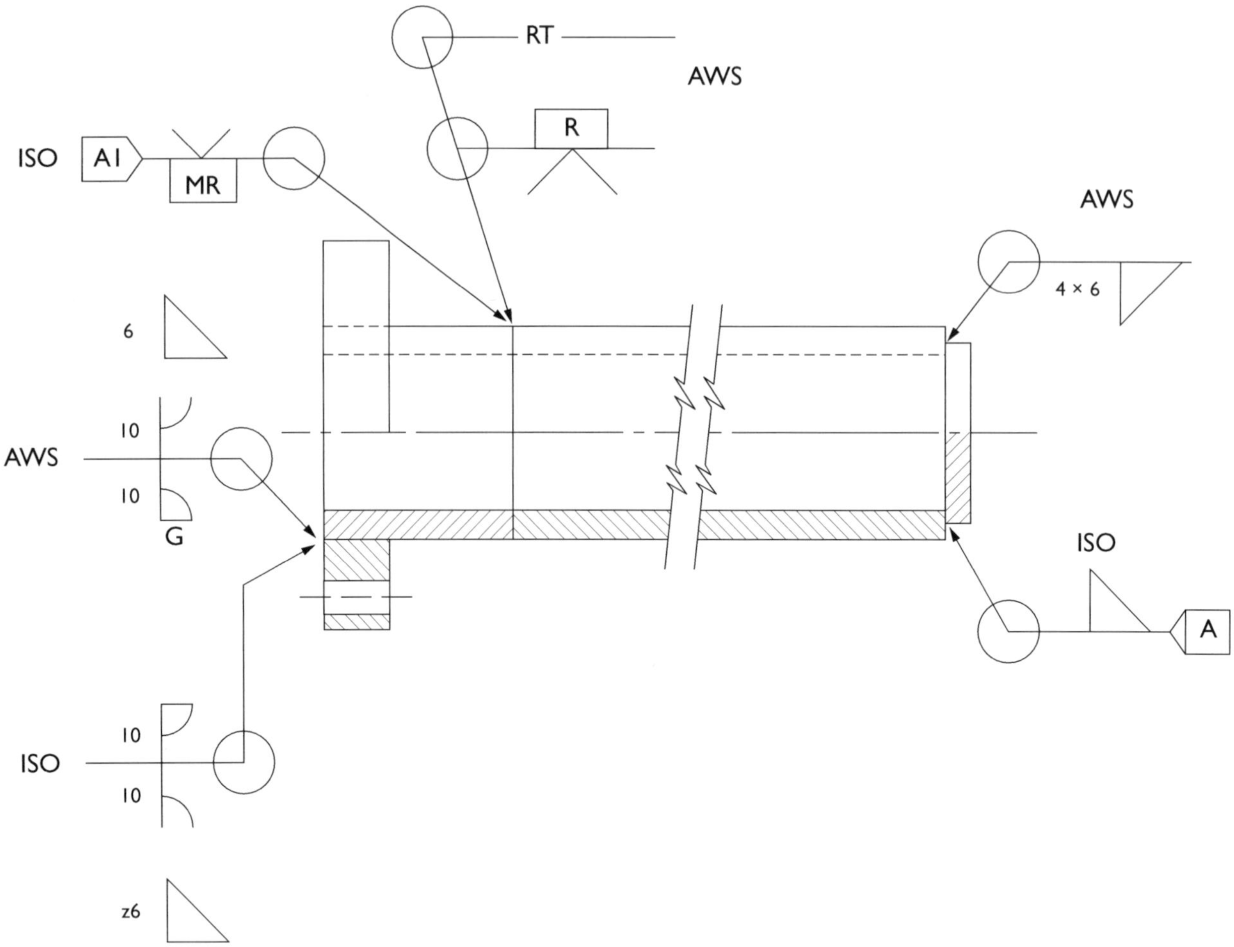

17.5 Exercise 1 solution (dimensions are in millimetres).

Comments

In **Fig. 17.5**, the AWS welding symbol for the circumferential butt weld includes an instruction for radiographic testing. There is no ISO symbol for NDT so the testing requirement is indicated by the A1 notation in the tail which refers to a note on the drawing or in the Welding Procedure Specification (WPS).

The weld on the face of the flange is to be finished flush with the end of the pipe face, which the AWS symbol indicates by the letter G for grinding. The ISO symbol implies that mechanical means of finishing will be required but the method is not specified. The ISO 6 × 4 mm fillet weld would be referred to in a separate drawing or instruction indicated by the letter A in the tail. In fact there is not much that the welder can do except fill the available space with weld metal.

Exercise 2 solution 1

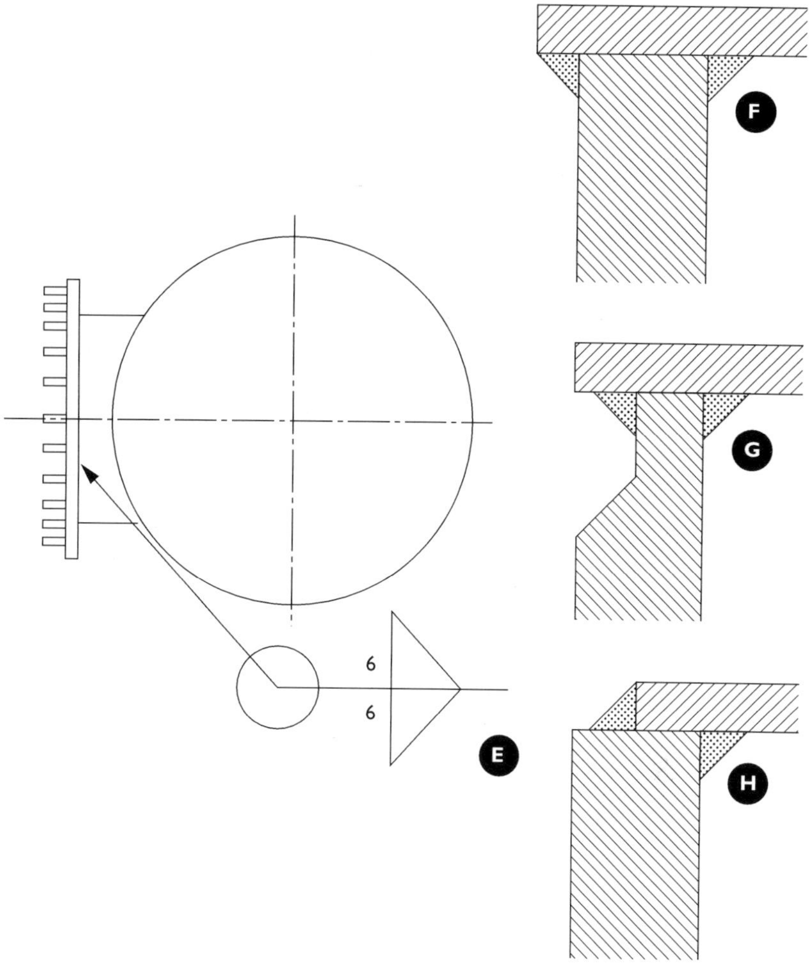

17.6 Exercise 2 solution 1 (dimensions are in millimetres).

Comments

In **Fig. 17.2** on page 45:

1. The standard used for the welding symbols has not been identified. There are clear indications that it is to ANSI/AWS.
   Ⓐ the process abbreviation in the fork is alphabetical and the weld symbol is for a stud weld.
   Ⓑ, Ⓒ and Ⓓ these are melt through symbols and the symbol for grinding.
   Ⓑ and Ⓓ these are radiographic test symbols.
2. Ⓑ and Ⓓ these symbols depict a single-V groove weld with melt through, ground flush inside and radiographed.
3. Ⓒ this symbol depicts a single-bevel-groove weld, stub pipe only bevelled (set-on branch, with melt through, ground flush inside).

Figure 17.6 shows the circled symbol of Ⓔ for ease of reference. This is a fillet weld, of 6 mm leg length, on both sides of a flange to tube joint.

Sketches F, G and H show possible interpretations of this joint. It is assumed that the studs would be welded to the flange before it is welded to the tube, in which case access for the double fillet weld in F and G would be severely restricted.

Option H is the most suitable design and it would be necessary to the fabricator to know the precise distance that the flange protrudes from the tube.

Exercise 2 solution 2

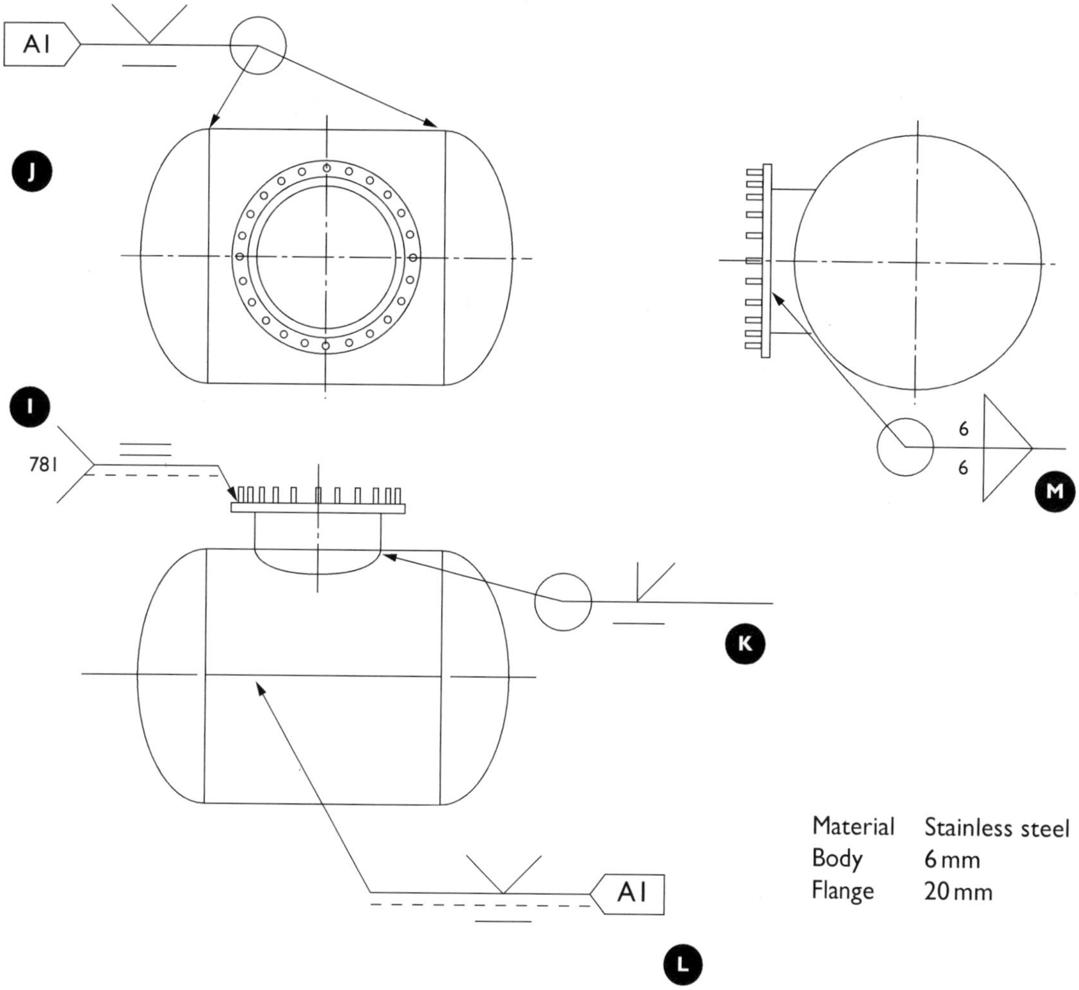

17.7 Exercise 2 solution 2 (dimensions are in millimetres).

Comments

In **Fig. 17.7**:

Ⓘ replaces Ⓐ in Fig. 17.2 on page 45. The weld symbol, consisting of two parallel lines, is designated a surface joint in ISO 2553, with an illustration that appears to represent a stud weld on a plate. The application of the symbol is assumed to be as shown.

Ⓙ replaces Ⓑ in Fig. 17.2 on page 45. Note that ISO 2553 does not have a symbol for NDT and uses a reference in the tail to indicate that instructions are given in a note on the drawing or in a Welding Procedure Specification (WPS). There is no melt through symbol in ISO 2553 and full penetration of a butt weld is always assumed, which would produce a penetration bead. The flat contour symbol is used by ISO without any indication of whether it is to be achieved in the as-welded condition or by any mechanical procedure. The reference in the tail, as well as indicating the NDT requirements, will contain instructions for any grinding procedure.

K and **L** replace **C** and **D** in Fig. 17.2 on page 45 (see comments on **J** above).

M this symbol is identical with the AWS symbol because the dashed reference line is omitted, which is standard practice for double butt or double fillet weld symbols.

Exercise 3 solution

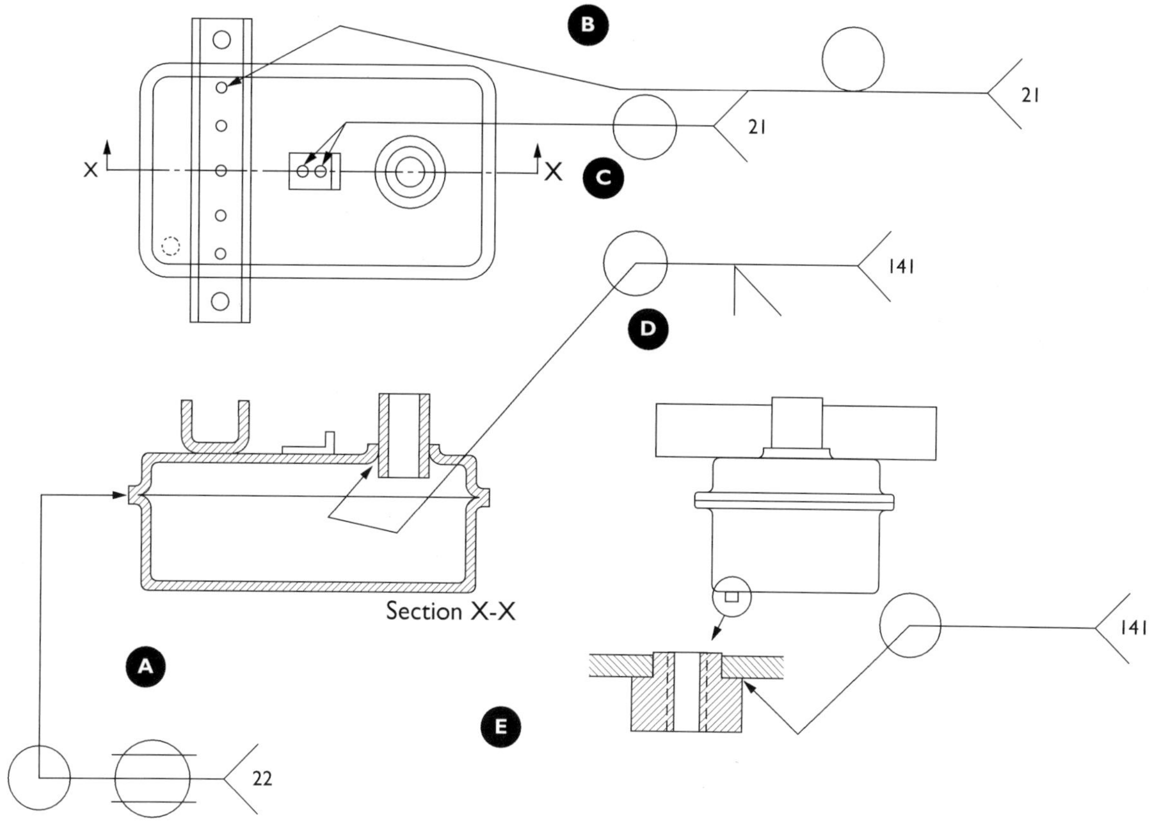

17.8 Exercise 3 solution.

Comments

In **Fig. 17.8**:

A shows welds joining walls of a tank. Gas welding, oxyfuel gas welding or TIG (GTAW) welding could be used here but, as the components are likely to be rough and not flush unless they are particularly good pressings, welding would be difficult on such thin sheet.

A better alternative is resistance seam welding which will tolerate components with mating surfaces that are not quite flat. It is important to note that the width of the flange must be large enough to allow access for the electrode wheels. The radius of the wheels must be less than the radius of the corners to prevent irregular contact with the consequent variation in weld quality.

B shows spot welds attaching a channel section. The obvious choice is resistance spot welding carried out before seam welding but there would then not be sufficient clearance for the seam welding electrode wheels. TIG (GTAW) spot welding is selected but the fabricator may suggest that this procedure even for a prototype job will involve considerable setting up time. A simpler procedure would be to attach the channel section by two short TIG (GTAW) fillet welds each side.

C shows spot welds attaching a bracket. Resistance spot welding before **A** is chosen, but arc spot welding would be equally suitable.

D shows a tube to tank top and **E** shows a capped boss to tank bottom. TIG (GTAW) welding is chosen but both joints would be ideal for brazing.

There are other possible choices of welding processes apart from those mentioned above, which emphasises the need for cooperation between the designer and fabricator. This cooperation is sometimes essential before welding symbols are added to a drawing, as mentioned previously.

Exercise 4 solution

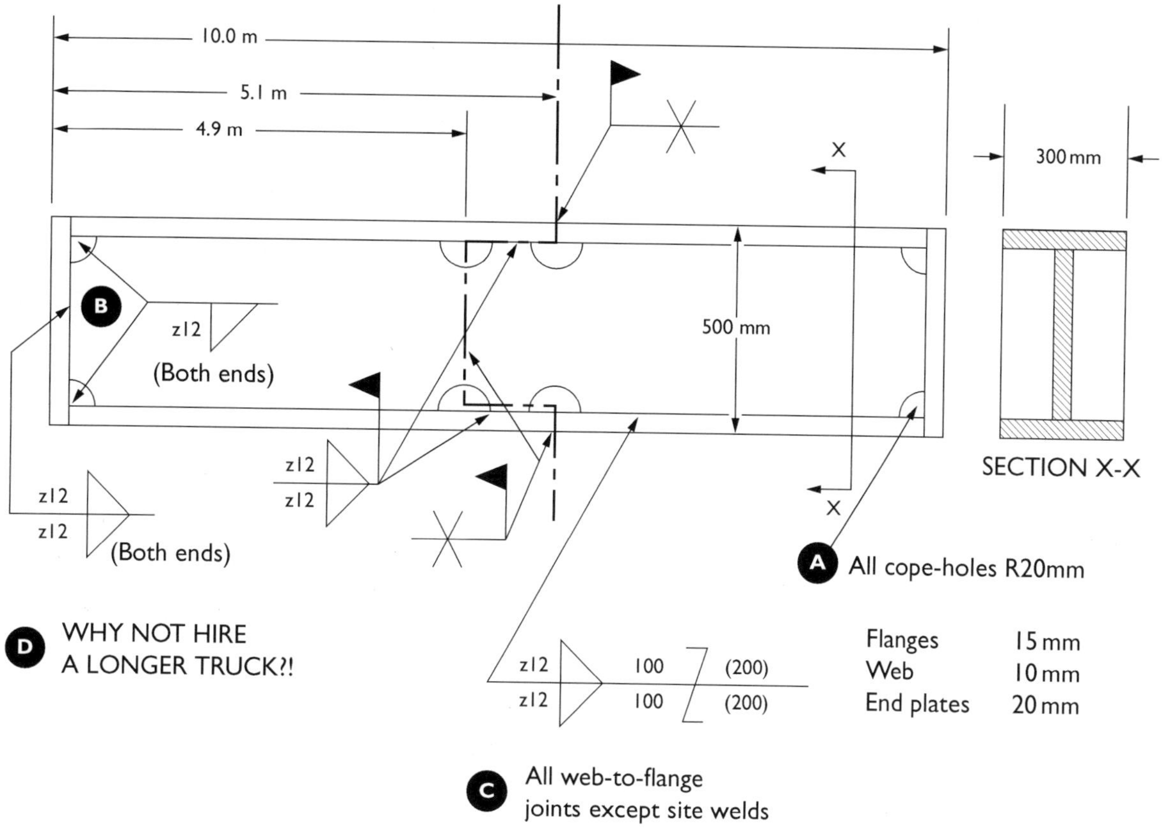

17.9 Exercise 4 solution. The dimensions shown in the welding symbols are in millimetres.

Comments
In **Fig. 17.9**:

A cope holes have been introduced to avoid the need to dress welds where they meet other welds and avoid welding up into corners, often a site for defects. Eight cope holes R20 mm are used.

B a weld all round symbol cannot be used for the end of the flange joints as it cannot go over the top and bottom of the flanges. Also, it is interrupted by the cope holes.

C there are four web to flange joints which would need four arrow lines if they were to be shown individually.

The spacing in ISO is not the pitch (here 300 mm) but the distance between the weld elements (here 300 – 100 = 200 mm).

D a suggestion to be taken seriously! Apart from simplifying the drawing, complete shop fabrication can reduce considerably the costs of welding and quality assurance.

However, as a 10 metre beam is being fabricated in two halves by manual welding, the fabricator may not have the space available or the handling facilities for a long beam and access from the fabrication shop or at the site maybe restricted. Again it is emphasised that cooperation between the design office and the fabrication shop and also the site office is essential.

Made in United States
Orlando, FL
04 November 2026

90200456R00039